2006

INTERNATIONAL PLUMBING CODE®

STUDY COMPANION

2006 International Plumbing Code

Study Companion

ISBN 978-1-58001-660-5

Cover Design:	Linda Cohoat
Publications Manager:	Mary Lou Luif
Project Editor:	Roger Mensink
Illustrator/Interior Design:	Mike Tamai
Manager of Development:	Doug Thornburg

First Printing: October 2007
Second Printing: February 2008

Printed in the United States of America

TABLE OF CONTENTS

INTRODUCTION

This study companion provides practical learning assignments for independent study of the provisions of the 2006 *International Plumbing Code®* (IPC®). The independent study format affords a method for the student to complete the program in an unregulated time period. Progressing through the workbook, the learner can measure his or her level of knowledge by using the exercises and quizzes provided for each study session.

The workbook is also valuable for instructor-led programs. In jurisdictional training sessions, community college classes, vocational training programs and other structured educational offerings, the study guide and the IPC can be the basis for classroom instruction.

All study sessions begin with a general learning objective specific to the session, the specific code sections or chapter under consideration and a list of questions summarizing the key points of study. Each session addresses selected topics from the IPC and includes code text, a commentary on the code provisions, illustrations representing the provisions under discussion and multiple choice questions that can be used to evaluate the student's knowledge. Before beginning the quizzes, the student should thoroughly review the IPC, focusing on the key points identified at the beginning of each study session.

The workbook is structured so that after every question the student has an opportunity to record his or her response and the corresponding code reference. The correct answers are found in the back of the workbook in the answer key.

Although this study companion is primarily focused on those subjects of specific interest to plumbing inspectors and contractors, it is a valuable resource to any individuals who would like to learn more about the IPC provisions. The information presented may be of importance to many building officials, plans examiners and combination inspectors.

The main responsibility for content development was with Ronald L. George, CIPE, CPD, President of Ron George Plumbing Design and Consulting Services. Special thanks are also extended to Joel Shelton, plumbing code instructor and consultant, for the review and development of quizzes; Steve Van Note, ICC Senior Technical Staff, for the review, organization and coordination of the content; and Hamid Naderi, ICC Vice-President of Product Development, for support and overall project management of this publication.

The information presented in this publication is believed to be accurate; however, it is provided for informational purposes only and is intended for use only as a guide. As there is a limited discussion of selected code provisions, the code itself should always be referenced for more complete information. In addition, the commentary set forth may not necessarily represent the views of any enforcing agency, as such agencies have the sole authority to render interpretations of the IPC.

Questions or comments concerning this study companion are encouraged. Please direct your comments to ICC at *studycompanion@iccsafe.org*.

Study Session

1

2006 IPC Sections 101 – 105
Administration I

OBJECTIVE: To develop an understanding of the purpose, intent and scope of the *International Plumbing Code*. To develop an understanding of the duties and responsibilities of the plumbing official and the department of plumbing inspections.

REFERENCE: Sections 101 through 105, 2006 *International Plumbing Code*

KEY POINTS:
- Which areas or activities do the code provisions of the 2006 *International Plumbing Code* cover?
- Which code regulates the installation of fuel gas distribution piping and equipment?
- When different sections of the code appear to be in conflict, which section applies?
- How are existing systems addressed by the code?
- What authority does the code official have in the maintenance of plumbing systems?
- How is the code official involved in a historic building?
- Are the referenced codes and standards usable as code provisions?
- What happens if there is a conflict between the code and the referenced standard or the manufacturer's installation instructions?
- Who is responsible for determining requirements not specifically covered by the code?
- Are there conditions under which a code official is liable?
- Under what conditions may code officials enter a building to inspect or perform their duty?
- What type of records are required to be kept by the code official?
- What are the conditions for approvals of alternatives?
- Under what conditions would a code official require tests to be made?
- Which conditions of approval are placed on an alternative engineered design?
- What information is required to be shown on construction documents?
- Are there conditions for approving the reuse of materials, equipment and devices?

Code Text: *These regulations shall be known as the* International Plumbing Code *of [NAME OF JURISDICTION] hereinafter referred to as "this code."*

Discussion and Commentary: This section sets forth the scope and intent of the code as it applies to new and existing structures. The adopted regulations are identified by inserting the name of the adopting jurisdiction into the code. This code is founded on principles intended to establish provisions consistent with the scope of a plumbing code that adequately protects public health, safety and welfare.

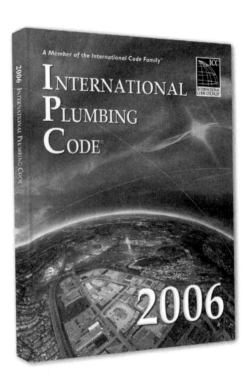

The IPC is intended to not unnecessarily increase construction costs; to not restrict the use of new materials, products or methods of construction; and to not give preferential treatment to particular types or classes of materials, products or methods of construction.

| **Topic:** Scope | **Category:** Administration |
| **Reference:** IPC 101.2 | **Subject:** General Provisions |

Code Text: *The provisions of* the International Plumbing Code *shall apply to the erection, installation, alteration, repairs, relocation, replacement, addition to, use or maintenance of plumbing systems within this jurisdiction.* The International Plumbing Code *shall also regulate nonflammable medical gas, inhalation anesthetic, vacuum piping, nonlexical oxygen systems and sanitary and condensate vacuum collection systems. The installation of fuel gas distribution piping and equipment, fuel gas-fired water heaters and water heater venting systems shall be regulated by* the International Fuel Gas Code. *Provisions in the appendices shall not apply unless specifically adopted.*

Exceptions:

1. *Detached one- and two-family dwellings and multiple single-family dwellings (townhouses) not more than three stories high with separate means of egress and their accessory structures shall comply with the* International Residential Code.

Discussion and Commentary: This section describes the types of plumbing system construction-related activities to which the code is intended to apply. The applicability of the code encompasses the initial design of plumbing systems, the installation and construction phases, and the maintenance of operating systems.

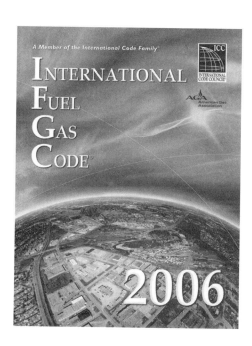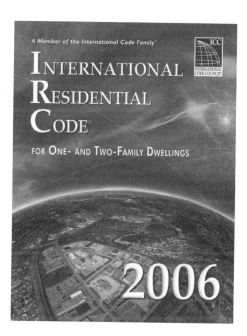

The *International Residential Code*® (IRC®) is a comprehensive code that contains provisions for all aspects of residential construction. The fuel gas portions of the IRC are based on the *International Fuel Gas Code*® (IFC®).

Code Text: *The purpose of* the International Plumbing Code *is to provide minimum standards to safeguard life or limb, health, property and public welfare by regulating and controlling the design, construction, installation, quality of materials, location, operation and maintenance or use of plumbing equipment and systems.*

Discussion and Commentary: The intent of the code is to set forth requirements that establish the minimum acceptable level to safeguard life or limb, health, property and public welfare. The intent becomes important in the application of such sections as:

- Section 102, Applicability
- Section 104.2, Rule-making authority
- Section 105.2, Alternative materials
- Section 108, Violations

These sections require enforcement-oriented interpretive action or judgment. As in any code, the written text is subject to interpretation. The inspector or authority having jurisdiction shall provide this interpretation.

Plumbing inspectors make application interpretations and judgment calls on a daily basis for installations that meet the intent of the code.

Topic: Existing installations

Category: Administration

Reference: IPC 102.2

Subject: Applicability

Code Text: *Plumbing systems lawfully in existence at the time of the adoption of this code shall be permitted to have their use and maintenance continued if the use, maintenance or repair is in accordance with the original design and no hazard to life, health or property is created by such plumbing system.*

Discussion and Commentary: The code is designed to regulate new construction and new work and is not intended to be applied retroactively to existing buildings, except where existing plumbing systems create unsafe conditions and are specifically addressed in this section and Section 108.

An existing building such as the above, which had its plumbing system designed and installed based on an older plumbing code, is not required to be brought up to code except for hazardous conditions.

Topic: Maintenance

Reference: IPC 102.3

Category: Administration

Subject: Applicability

Code Text: *All plumbing systems, materials and appurtenances, both existing and new, and all parts thereof, shall be maintained in proper operating condition in accordance with the original design in a safe and sanitary condition. All devices or safeguards required by this code shall be maintained in compliance with the code edition under which they were installed.*

The owner or the owner's designated agent shall be responsible for maintenance of plumbing systems. To determine compliance with this provision, the code official shall have the authority to require any plumbing system to be reinspected.

Discussion and Commentary: All plumbing systems and equipment are subject to deterioration resulting from aging, wear, accumulation of dirt and debris, corrosion and other factors. Maintenance is necessary to keep plumbing systems and equipment in proper operating condition.

A plumber's maintenance truck should contain all the necessary supplies and materials needed to keep plumbing systems and equipment in proper operating condition. The owner is responsible for the proper working condition of the system.

Topic: Addition, Alterations or Repairs

Category: Administration

Reference: IPC 102.4

Subject: Applicability

Code Text: *Additions, alterations, renovations or repairs to any plumbing system shall conform to that required for a new plumbing system without requiring the existing plumbing system to comply with all the requirements of this code. Additions, alterations or repairs shall not cause an existing system to become unsafe, insanitary or overloaded.*

Minor additions, alterations, renovations and repairs to existing plumbing systems shall be permitted in the same manner and arrangement as in the existing system, provided that such repairs or replacement are not hazardous and are approved.

Discussion and Commentary: The intent of this section is to allow the continued use of existing plumbing systems and equipment that may not be designed and constructed as required for new installations. Existing plumbing systems and equipment will normally require repair and component replacement to remain operational. This section permits repair and component replacements to occur without requiring the redesign, alteration or replacement of the entire system.

Many elements of the plumbing system will require repair or are subject to alteration or addition during the building lifetime. Depending on the extent of such repairs, alterations or additions, the currently adopted plumbing provisions might apply.

Topic: Historic Buildings

Reference: IPC 102.6

Category: Administration

Subject: Applicability

Code Text: *The provisions of the International Plumbing Code relating to the construction, alteration, repair, enlargement, restoration, relocation or moving of buildings or structures shall not be mandatory for existing buildings or structures identified and classified by the state or local jurisdiction as historic buildings when such buildings or structures are judged by the code official to be safe and in the public interest of health, safety and welfare regarding any proposed construction, alteration, repair, enlargement, restoration, relocation or moving of buildings.*

Discussion and Commentary: This section provides the code official with the widest possible flexibility in enforcing the code when the building in question has historic value. This flexibility, however, is not provided without conditions. The most important criterion for application of this section is that the building must be specifically classified as being of historic significance by a qualified party or agency.

The most important issue for consideration regarding historic buildings is whether the proposed plumbing work will alter or destroy the historic elements or features of the building.

Topic: Referenced Codes and Standards   **Category:** Administration
Reference: IPC 102.8   **Subject:** Applicability

Code Text: *The codes and standards referenced in the International Plumbing Code shall be those that are listed in Chapter 13 and such codes and standards shall be considered as part of the requirements of this code to the prescribed extent of each such reference. Where the differences occur between provisions of this code and the referenced standards, the provisions of this code shall be the minimum requirements.*

Discussion and Commentary: The code references many standards promulgated and published by other organizations. A complete list of referenced standards appears in Chapter 13. The wording of this provision ("shall be those that are listed in Chapter 13") was carefully chosen in order to establish the edition of the standard that is enforceable under the code. The reference to a standard means that only the portions pertaining to the IPC provision regulating a specific subject are adopted, and not the entire standard.

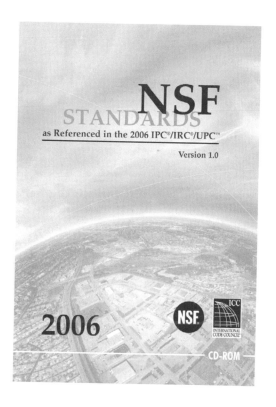

There are nine NSF Standards referenced in the 2006 IPC. Other standards such as ASME, ACSE and ASTM are also extensively referenced. In case of a conflict between the IPC and a referenced standard, the code provisions prevail.

Topic: Requirements Not Covered by the IPC   **Category:** Administration

Reference: IPC 102.9   **Subject:** Applicability

Code Text: *Any requirements necessary for the strength, stability or proper operation of an existing or proposed plumbing system, or for the public safety, health and general welfare, not specifically covered by* the International Plumbing Code *shall be determined by the code official.*

Discussion and Commentary: Evolving technology in our society will inevitably result in a situation or circumstance in which the code is comparatively silent on an identified hazard. The reasonable application of the code to such hazardous, unforeseen conditions is provided in this section. Clearly, such a section is needed, and the code official's judicious and reasonable application of the section is necessary.

When the code is silent pertaining to new technology, it is the code official's responsibility to determine the safety, health and general welfare pertaining to any newly installed equipment or systems. The code is not intended to prevent or stifle new technologies or innovative systems.

Topic: Appointment	**Category:** Administration
Reference: IPC 103.2	**Subject:** Department of Plumbing Inspection

Code Text: *The code official shall be appointed by the chief appointing authority of the jurisdiction, and the code official shall not be removed from office except for cause and after full opportunity to be heard on specific and relevant charges by and before the appointing authority.*

Discussion and Commentary: This section establishes the code official as an appointment position from which the official cannot be removed, except for cause subject to a due process review. Many jurisdictions have their own specific administrative rules regarding the employment and separation of employees and might therefore amend this section of the IPC.

CODE OFFICIAL: The officer or other designated authority charged with the administration and enforcement of this code, or a duly authorized representative.

Code Text: *The code official shall have authority as necessary in the interest of public health, safety and general welfare to adopt and promulgate rules and regulations to interpret and implement the provisions of this code to secure the intent thereof and to designate requirements applicable because of local climatic or other conditions. Such rules shall not have the effect of waiving structural or fire performance requirements specifically provided for in this code, or of violating accepted engineering practice involving public safety.*

Discussion and Commentary: The code official has the administrative authority to promulgate rules and regulations that serve to interpret or supplement the provisions of the code, as long as such rules conform to the intent of the code. Most important, these rules are not intended to set aside or waive any code provisions, but are intended to facilitate code compliance. The most frequent use of this authority is the promulgation of the administrative rules and procedures for the department's efficient and effective operation.

The code official must ensure the public safety by developing rules and procedures to enforce the jurisdiction's adopted codes and by ensuring that accepted engineering practices involving public safety are not violated.

Code Text: *The design, documentation, inspection, testing and approval of an alternative engineered design plumbing system shall comply with Sections 105.4.1 through 105.4.6.*

Discussion and Commentary: This section provides the opportunity for an engineer or architect to design a plumbing system that may not comply with all of the provisions found in Chapters 3 through 13. The design must be approved by the code official and must conform to accepted engineering principles. The engineered plumbing system must provide the level of protection of the public health, safety and welfare intended by the IPC.

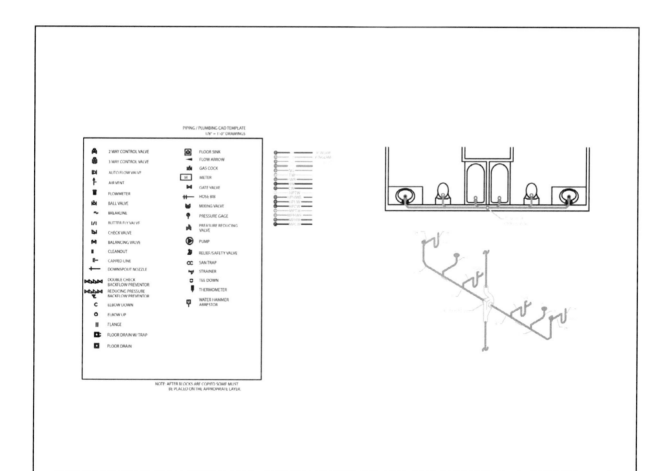

IPC Section 15.4.2 further requires that the registered design professional must indicate on the permit application that the plumbing system is an alternative engineered design. When future permits regarding alterations or modifications are applied for, appropriate measures can then be taken to determine that the future work will not adversely affect the system design.

Topic: Material and Equipment Reuse

Category: Administration

Reference: IPC 105.5

Subject: Approval

Code Text: *Materials, equipment and devices shall not be reused unless such elements have been reconditioned, tested, placed in good and proper working condition and approved.*

Discussion and Commentary: The code criteria for materials and equipment have changed over the years. Evaluation of testing and materials technology has permitted the development of new criteria, which the old materials may not satisfy. As a result, used materials are required to be evaluated in the same manner as new materials. Used (previously installed) equipment must be equivalent to that required by the code if it is to be utilized in a new installation.

Reconditioned materials and devices are acceptable if approved. **APPROVED:** Acceptable to the code official or other authority having jurisdiction.

Quiz

Study Session 1
IPC Sections 101 – 105

1. The installation of a gas-fired water heater in a school building is regulated by the *International Plumbing Code* and the _____.

    a. *International Building Code*® (IBC®)

    b. *International Fuel Gas Code*® (IFGC®)

    c. *International Fire Code*® (IFC®)

    d. *International Mechanical Code*® (IMC)®

    Reference _____

2. The repair of plumbing in a two-story townhouse is regulated by which of the following codes?

    a. *International Plumbing Code*® (IPC®)

    b. *International Existing Building Code*® (IEBC®)

    c. *International Residential Code*® (IRC®)

    d. *International Property Maintenance Code*® (IPMC®)

    Reference _____

3. The provisions of the appendices do not apply unless _____.

    a. referenced in the code

    b. applicable to specific conditions

    c. specifically adopted

    d. relevant to fire or life safety

    Reference _____

4. The pipe sizing tables for a fuel gas water heater in a three-story apartment complex are found in the _____.

    a. IFGC

    b. IMC

    c. IRC

    d. manufacturer's installation instructions

    Reference _____

5. Which of the following is *not* within the scope of the IPC?

    a. nonflammable medical gas systems

    b. drinking water treatment systems

    c. private sewage disposal systems

    d. vacuum piping systems

    Reference _____

6. In order to safeguard life, health and property, the IPC regulates all of the following within the built environment, except _____.

    a. installation and use   b. design and maintenance

    c. quality of material   d. cost of construction

    Reference _____

7. Where differences occur between specific requirements in different sections of the IPC, the _____ requirement shall apply.

    a. general   b. most restrictive

    c. least restrictive   d. specific

Reference _____

8. Existing plumbing systems must be brought into compliance with the IPC if the system is _____.

    a. repaired   b. added on to

    c. a hazard to property   d. nonconforming

Reference _____

9. The _____ shall be responsible for the maintenance of the plumbing system.

    a. maintenance staff

    b. person occupying the structure

    c. installing contractor

    d. owner or owner's agent

Reference _____

10. Where code provisions are in conflict with the referenced standard, the _____ shall apply.

    a. code provisions   b. referenced standard

    c. most restrictive   d. specific requirement

Reference _____

11. When requirements necessary for the proper operation of a proposed plumbing system are not specifically covered by the code, such requirements shall be determined by _____.

    a. a registered design professional

    b. a licensed contractor

    c. the board of appeals

    d. the code official

Reference _____

12. The code official has authority to _____ the provisions of the code.

    a. ignore   b. waive

    c. violate   d. interpret

Reference _____

13. The code official has authority to grant modifications to the code _____ .

    a. for only those issues not involving fire or life safety

    b. for individual cases where the strict letter of the code is impractical

    c. where the intent and purpose of the code cannot be met

    d. related only to the administrative functions

Reference _____

14. Tests performed by _____ may be required by the code official where there is insufficient evidence of code compliance.

    a. the owner   b. the contractor

    c. an approved agency   d. a design professions

Reference _____

15. In order for an alternative material, design or method of construction to be considered acceptable, it must be equivalent to the code based on all but which of the following criteria:

    a. durability   b. economics

    c. strength   d. fire resistance

Reference _____

16. The _____ shall indicate on the permit application that the plumbing system is an alternative engineered design.

    a. registered design professional

    b. code official

    c. owner

    d. contractor

Reference _____

17. Plumbing equipment may be reused only when _____ .

    a. in good condition
    b. approved by the code official

    c. it bears the original label
    d. approved by the manufacturer

Reference _____

18. Minor additions and alterations to an existing plumbing system may match the installation methods of the existing system if _____ .

    a. in compliance with the intent of the code

    b. approved by the code official

    c. they do not overload the system

    d. sanitary

Reference _____

19. Required department records shall be maintained _____ .

    a. permanently

    b. as required by the code official

    c. for as long as the building is in existence

    d. for 5 years after construction is complete, subject to other regulations

Reference _____

20. Provisions of this code are not mandated for historic buildings _____ and judged safe by the code official.

    a. classified as historic by the state

    b. constructed prior to adoption of a plumbing code

    c. located within the boundaries of a designated historic district

    d. as approved by the board of appeals

Reference _____

21. After issuance of a permit, the code official is authorized to_____.

    a. waive required inspections

    b. enter the building at any time

    c. accept written inspection reports

    d. discard inspection reports once construction is complete

Reference _____

22. The code official shall maintain records of all of the following except _____.

    a. notices and orders   b. applications

    c. fees collected   d. construction documents

Reference _____

23. The code official shall be appointed by the _____.

    a. chief appointing authority

    b. building official

    c. director of the plumbing inspection department

    d. jurisdiction attorney

Reference _____

24. The prescribed portions of standards referenced in the IPC and listed in Chapter 13 are considered _____.

    a. requirements   b. guidelines

    c. alternatives   d. modifications

Reference _____

25. The registered design professional must submit _____ to substantiate that the performance of an alternative engineered design meets the intent of the code.

    a. construction documents

    b. design criteria

    c. manufacturer's installation instructions

    d. technical data

Reference _____

2006 IPC Sections 106 – 109
Administration II

OBJECTIVE: To gain an understanding of the required permits, permit procedures, fees and violations. To develop an understanding of the appeals board, its membership and responsibilities.

REFERENCE: Sections 106 through 109, 2006 *International Plumbing Code*

KEY POINTS:
- Under what conditions are plumbing permits required?
- What is required on the permit application?
- When are penetrations required to be identified for fire blocking, fire-resistance ratings, and structural safety?
- Under what circumstances would a permit become null and void?
- How many and what are the required inspections?
- What type of system is required to have special inspections?
- Under what condition may the code official allow a temporary connection?
- What is the penalty for continuing to work after a stop work order is issued?
- Under what conditions should a code official condemn equipment?
- What two conditions would justify the disconnection of utilities?
- Who may attend hearings before the board of appeals?
- How many votes are required to modify or reverse the code official's decision?

Code Text: *Each application for a permit, with the required fee, shall be filed with the code official on a form furnished for that purpose and shall contain a general description of the proposed work and its location. The application shall be signed by the owner or and authorized agent. The permit application shall indicate the proposed occupancy of all parts of the building and of that portion of the site or lot, if any, not covered by the building or structure and shall contain such other information required by the code official.*

Discussion and Commentary: This section limits persons who may apply for a permit to the building owner or an authorized agent. An owner's authorized agent could be anyone who is given written permission to act in the owner's interest for the purpose of obtaining a permit, such as an architect, engineer, contractor, tenant or other. Permit forms generally have sufficient space to write a very brief description of the work to be accomplished, which is acceptable for small jobs. For larger projects, the description will be augmented by construction documents.

SAMPLE
PLUMBING PERMIT FEE SCHEDULE/APPLICATION

Date: _____
Contractor: _____
Address: _____
City: _____
Phone#: _____
License: _____
Federal Employee I.D.#: _____
Employee #: _____
Worker's Comp Insurance: _____

Receipt #: _____
Owner: _____
Owner's Address: _____
Owner's City: _____
Owner's Phone: _____

Type of Job (Check all that apply)
Commercial   Residential   Industrial
New Const.   Remodel   Replacement

Item Description	Fee ($)	# of Items	Total
All permits are subject to an administrative fee of:	$30.00		
New Home Construction: Single family Dwelling up to 2 inspections (one rough and one final)	$200.00		
Per Scheduled Home Inspection (During normal business hours)	$75.00		
Each Additional bathroom after two	$30.00		
Underground Inspection	$32.00		
Mobile Home (final only)	$75.00		
Modular Homes (final only)	$150.00		
Commercial Base Fee/Plan Review	$80.00		
Reduced Pressure Backflow Preventor/Backwater Valve	$9.00		
Stacks, Vents, Conductors	$9.00		
Drains: floor, roof.	$9.00		
Interceptor, grease, oil and other.	$9.00		
Sewer: sanitary, storm, gas system, soil & waste re-piping	$9.00		
Disposal, Dishwasher, Refrigerator & Washing Machine	$9.00		
Water Heater, Water Softener, Sump Pump	$9.00		
Ice-making Machine, Drinking Fountain	$9.00		
Heating System make-up water connection	$9.00		
Toilet, Bathtub, Shower Stall & Sink or Slop Sink	$9.00		
Lavatories & Urinal	$9.00		
Sill Cocks & Laundry Tub	$9.00		
Lawn Sprinkler System	$30.00		
Other Inspections and Fees:			
Inspection outside of normal business hours (minimun charge two hours)	$90.00 per hour		
Reinspection fee assessed under provisions of Chapter 1 Reinspection and Testing Section 107.3.3	$75.00 each		
Inspections for which no fee is specifically indicated (minimum charge of one-half hour)	$75.00 per hour		
Additional plan review required by changes, additions or revisions to approved plans (minimum charge one-half hour	$75.00 per hour		

_____
Signature of Applicant

Each jurisdiction can develop its own permit application forms and fee schedules to be compatible with its jurisdictional rules, policies and requirements.

Code Text: *Every permit issued by the code official under the provisions of this code shall expire by limitation and become null and void if the work authorized by such permit is not commenced within 180 days from the date of such permit, or if the work authorized by such permit is suspended or abandoned at any time after the work is commenced for a period of 180 days. Before such work can be recommenced, a new permit shall be first obtained and the fee therefore shall be one-half the amount required for a new permit for such work, provided no changes have been made or will be made in the original construction documents for such work, and provided further that such suspension or abandonment has not exceeded 1 year.*

Discussion and Commentary: The permit becomes invalid under two distinct situations, both based on a six-month period. The first situation exists when no work has started six months from issuance of the permit. The second situation exists when there is no continuation of authorized work for six months. The person who was issued the permit should be notified in writing that it is invalid and what steps must be taken to restart the work.

January 2008	Sun	Mon	Tue	Wed	Thu	Fri	Sat
			1	2	3	4	5
	6	7	**8**	9	10	11	12
	13	14	15	16	17	18	19
	20	21	22	23	24	25	26
	27	28	29	30	31		

July 2008	Sun	Mon	Tue	Wed	Thu	Fri	Sat
			1	2	3	4	5
	6	7	**8**	9	10	11	12
	13	14	15	16	17	18	19
	20	21	22	23	24	25	26
	27	28	29	30	31		

A regularly updated manual or automated reporting system must be developed to aid the code official in identifying which permits have been inactive within certain time periods so they can be followed for compliance with Section 106.5.3.

Code Text: *A permit shall not be issued until the fees prescribed in Section 106.6.2 have been paid, and an amendment to a permit shall not be released until the additional fee, if any, due to an increase of the plumbing systems, has been paid.*

Discussion and Commentary: All fees are to be paid prior to permit issuance. This requirement both establishes that the permit applicant intends to proceed with the work and facilitates payment. Fees are typically developed by jurisdictions to cover the costs associated with the departments' activities to ensure installations are in compliance with the codes. A sample fee schedule is provided in Appendix A of the IPC.

APPENDIX A

PLUMBING PERMIT FEE SCHEDULE

Permit Issuance

1. For issuing each permit . . . . $ _____
2. For issuing each supplemental permit . . . . _____

Unit Fee Schedule

1. For each plumbing fixture or trap or set of fixtures on one trap (including water, drainage piping and backflow protection thereof) . . . . _____
2. For each building sewer and each trailer park sewer . . . . _____
3. Rainwater systems—per drain (inside building) . . . . _____
4. For each cesspool (where permitted) . . . . _____
5. For each private sewage disposal system . . . . _____
6. For each water heater and/or vent . . . . _____
7. For each industrial waste pretreatment interceptor including its trap and vent, excepting kitchen-type grease interceptors functioning as fixture traps . . . . _____
8. For installation, alteration or repair of water-piping and/or water-treating equipment, each . . . . _____
9. For repair or alteration of drainage or vent piping, each fixture . . . . _____
10. For each lawn sprinkler system on any one meter including backflow protection devices therefor . . . . _____
11. For atmospheric-type vacuum breakers not included in Item 2:

    1 to 5 . . . . _____

    over 5, each . . . . _____
12. For each backflow protective device other than atmospheric-type vacuum breakers:

    2 inches (51 mm) and smaller . . . . _____

    Over 2 inches (51 mm) . . . . _____

Other Inspections and Fees

1. Inspections outside of normal business hours . . . . _____ per hour
   (minimum charge two hours)
2. Reinspection fee assessed under provisions of Section 107.3.3 . . . . _____ each
3. Inspections for which no fee is specifically indicated . . . . _____ per hour
   (minimum charge one-half hour)
4. Additional plan review required by changes, additions or revisions to approved plans (minimum charge one-half hour) . . . . _____ per hour

---

Many departments have developed flexible payment procedures such as an active account for the applicant, similar to a bank account from which permit fees are deducted every time an application is filed. Payment by credit card is also commonly allowed.

Topic: Work Commencing before Permit   **Category:** Administration
Reference: IPC 106.6.1   **Subject:** Permits

Code Text: *Any person who commences any work on a plumbing system before obtaining the necessary permits shall be subject to 100 percent of the usual permit fee in addition to the required permit fees.*

Discussion and Commentary: This section is intended to serve as a deterrent to proceeding with work on a plumbing system without a permit (except as provided in Section 106.2). As a punitive measure, it doubles the cost of the permit fee. This section does not, however, intend to penalize a contractor called upon to do emergency work after hours, provided that he or she makes prompt notification to the code official the next business day and obtains the requisite permit for the work done and has the required inspections performed.

Sometimes referred to as an *investigation fee*, the additional fee is intended to cover the additional costs associated with the time and investigation that takes place to identify construction activities that go on without a permit and to make proper inspection of what had already been installed without prior approval.

Code Text: *The fees for all plumbing work shall be as indicated in the following schedule: [JURISDICTION TO INSERT APPROPRIATE SCHEDULE]*

Discussion and Commentary: A published fee schedule must be established for plans examination, permits and inspections. Ideally, the department should generate revenues that cover operating costs and expenses. The permit fee schedule is an integral part of this process. Many jurisdictions have a book of fees or a fee ordinance that shows the schedule of fees for all jurisdictional services including permit fees.

Plumbing fees are sometimes combined with other fees related to the same project. Sometimes referred to as *project permits*, such permits cover all aspects of a project, such as plumbing, mechanical, electrical and general building construction, with a project permit fee.

Code Text: *The code official, upon notification from the permit holder or the permit holder's agent, shall make the following inspections and such other inspections as necessary, and shall either release that portion of the construction or shall notify the permit holder or an agent of any violations that must be corrected. The holder of the permit shall be responsible for the scheduling of such inspections.*

> *1. Underground inspection shall be made after trenches or ditches are excavated and bedded, piping installed, and before any backfill is put in place.*
> *2. Rough-in inspection shall be made after the roof, framing, fireblocking, firestopping, draftstopping and bracing is in place and all sanitary, storm and water distribution piping is roughed-in, and prior to the installation of wall or ceiling membranes.*
> *3. Final inspection shall be made after the building is complete, all plumbing fixtures are in place and properly connected, and the structure is ready for occupancy.*

Discussion and Commentary: Inspections are necessary to determine that an installation conforms to all code requirements. Because the majority of a plumbing system is hidden within the building enclosure, periodic inspections are necessary before portions of the system are concealed. All inspections that are necessary to provide such verification must be conducted.

The code official is required to determine that plumbing systems and equipment are installed in accordance with the approved construction documents and the applicable code requirements.

Code Text: *New plumbing systems and parts of existing systems that have been altered, extended or repaired shall be tested as prescribed herein to disclose leaks and defects, except that testing is not required in the following cases:*

1. *In any case that does not include addition to, replacement, alteration or relocation of any water supply, drainage or vent piping.*
2. *In any case where plumbing equipment is set up temporarily for exhibition purposes.*

Discussion and Commentary: Every plumbing system must be tested before it is placed into service. Testing is necessary to make sure that the system is free from leaks or other defects. Testing is also required, to the extent practicable, for portions of existing systems that have been repaired, altered or extended.

Because plumbing pipes and components are for most part concealed within the building construction, any leaks could cause substantial expenses for the owner and interruptions for the tenants.

Code Text: *After the prescribed tests and inspections indicate that the work complies in all respects with the International Plumbing Code, a notice of approval shall be issued by the code official.*

Discussion and Commentary: After the code official has performed the required inspections and observed the required equipment and system tests (or has received written reports of the results of such tests), he or she must determine if the installation or work is in compliance with all applicable sections of the code. The code official must issue a written notice of approval if it has been determined that the subject plumbing work or installation is in apparent compliance with the code. The notice of approval is given to the permit holder, and a copy of such notice is retained on file by the code official.

Anytown Building Department

(123) 555-4567

INSPECTION APPROVED

☐ Building ☐ Plumbing

☐ Electrical ☐ Mechanical

Description: _____

Comments _____

_____

_____

Date: _____

Inspector: _____

This approval tag is referred to as a *green tag* by many jurisdictions because traditionally it has been coded green to signify the contractor may proceed. Because many approval tags could still contain minor corrective instructions, some jurisdictions have moved away from the color scheme for tags.

Topic: Notice of Violation

Category: Administration

Reference: IPC 108.2

Subject: Violations

Code Text: *The code official shall serve a notice of violation or order to the person responsible for the erection, installation, alteration, extension, repair, removal or demolition of plumbing work in violation of the provisions of this code, or in violation of a detail statement or the approved construction documents there under, or in violation of a permit or certificate issued under the provisions of this code. Such order shall direct the discontinuance of the illegal action or condition and the abatement of the violation.*

Discussion and Commentary: The code official is required to notify the person responsible for the erection or use of a building found to be in violation of the code. The section that is allegedly being violated must be cited so that the responsible party can respond to the notice.

Anytown Building Department

CORRECTION NOTICE

TO _____

ADDRESS _____

DESCRIPTION _____

The following work is not in compliance with the Code and requires correction:

_____

_____

_____

_____

_____

_____

_____

Work shall not be covered until inspected and approved by this department. When corrections are completed, call for inspection at (123) 555-4567.

_____   _____
INSPECTOR   DATE

Building Department
222 1st Street SW, Anytown
(123) 555-4567

Contractors are always free to discuss the failed inspection as well as the reasons behind code violations for items on which they might not agree with the inspector. They may also appeal the decision through the departmental chain of command.

Topic: Stop Work Orders

Category: Administration

Reference: IPC 108.5

Subject: Violations

Code Text: *Upon notice from the code official, work on any plumbing system that is being done contrary to the provisions of this code or in a dangerous or unsafe manner shall immediately cease. Such notice shall be in writing and shall be given to the owner of the property, or to the owner's agent, or to the person doing the work. The notice shall state the conditions under which work is authorized to resume. Where an emergency exists, the code official shall not be required to give a written notice prior to stopping the work. Any person who shall continue any work in or about the structure after having been served with a stop work order, except such work as that person is directed to perform to remove a violation or unsafe condition, shall be liable to a fine of not less than [AMOUNT] dollars or more than [AMOUNT] dollars.*

Discussion and Commentary: A stop work order can prevent a violation from becoming worse and more difficult or expensive to correct. A stop work order may be issued where work is proceeding without a permit to perform the work. Hazardous conditions can develop when the code official is unaware of the nature of the work and a permit for the work has not been issued.

Anytown Building Department

STOP WORK

TO _____

ADDRESS _____

DESCRIPTION _____

NOTICE

I HAVE THIS DAY INSPECTED THIS STRUCTURE AND THESE PREMISES AND HAVE FOUND THE FOLLOWING VIOLATIONS OF COUNTY LAWS GOVERNING SAME:

_____

_____

_____

_____

_____

_____

_____

WORK SHALL NOT RESUME UNTIL VIOLATION IS CORRECTED AND AUTHORIZATION TO PROCEED IS GRANTED.

_____ _____

INSPECTOR   DATE

DO NOT REMOVE THIS NOTICE

Building Department
222 1st Street SW, Anytown
(123) 555-4567

It is always a good idea to make telephone contact with the owners or contractors to inform them of the stop work order because the stop work tags sometimes get lost or blown away by wind.

Topic: Unsafe Plumbing

Reference: IPC 108.7

Category: Administration

Subject: Violations

Code Text: *Any plumbing regulated by this code that is unsafe or that constitutes a fire or health hazard, insanitary condition, or is otherwise dangerous to human life is hereby declared unsafe. Any use of plumbing regulated by this code constituting a hazard to safety, health or public welfare by reason of inadequate maintenance, dilapidation, obsolescence, fire hazard, disaster, damage or abandonment is hereby declared an unsafe use. Any such unsafe equipment is hereby declared to be a public nuisance and shall be abated by repair, rehabilitation, demolition or removal.*

Discussion and Commentary: Unsafe conditions include those that constitute a health hazard, fire hazard, explosion hazard, shock hazard, asphyxiation hazard, physical injury hazard or are otherwise dangerous to human life and property. In the course of performing duties, the code official may identify a hazardous condition. Such condition must be declared in violation of the code and must therefore be abated.

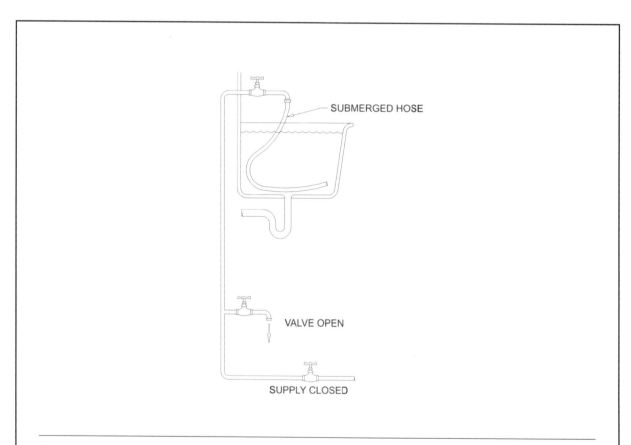

Sometimes while inspecting a plumbing installation, the code official may notice a leaking soil pipe that is causing structural decay, or a missing cleanout plug or defective water-heater relief valve. Though the defective plumbing may be unrelated to the installation that the code official was called to inspect, he or she is obligated to address the hazard and have it corrected.

Topic: Authority to Condemn Equipment

Category: Administration

Reference: IPC 108.7.1

Subject: Violations

Code Text: *Whenever the code official determines that any plumbing, or portion thereof, regulated by this code has become hazardous to life, health or property or has become insanitary, the code official shall order in writing that such plumbing either be removed or restored to a safe or sanitary condition. A time limit for compliance with such order shall be specified in the written notice. No person shall use or maintain defective plumbing after receiving such notice. When such plumbing is to be disconnected, written notice as prescribed in Section 108.2 shall be given. In cases of immediate danger to life or property, such disconnection shall be made immediately without such notice.*

Discussion and Commentary: Disconnection of a plumbing system from the utility supply is the most radical method of hazard abatement available to the code official and should be reserved for cases in which all other lesser remedies have proven ineffective. Such an action must be preceded by written notice to the owner and any occupants of the building being ordered to disconnect. Disconnection must be accomplished within the time frame established by the code official in the written notification of disconnect.

When the hazard to public health and welfare is so imminent as to mandate immediate disconnection, the code official has the authority and even the obligation to cause disconnection before following up with notice.

Code Text: *Any person shall have the right to appeal a decision of the code official to the board of appeals. An application for appeal shall be based on a claim that the true intent of this code or the rules legally adopted there under have been incorrectly interpreted, the provisions of this code do not fully apply, or an equally good or better form of construction is proposed. The application shall be filed on a form obtained from the code official within 20 days after the notice was served.*

Discussion and Commentary: This section literally allows any person to appeal a decision of the code official. In practice, this section has been interpreted to permit appeals only by those aggrieved parties with a material or definitive interest in the decision of the code official. An aggrieved party may not appeal a code requirement per se. The intent of the appeal process is not to waive or set aside the code requirement; rather, it is intended to provide a means of reviewing a code official's decision on an interpretation or application of the code, or to review the equivalency of protection to the code requirements.

CITY OF ANYTOWN
BUILDING DEPARTMENT
PLUMBING DIVISION

PLUMBING BOARD OF APPEALS

Appeal for interpretation of the Plumbing Code

Project address _____

Type of construction _____

Building use _____

Owner's name _____ Phone _____

Owner's address _____

In accordance with the provisions of the City of Anytown Plumbing Regulations Section 5.2.3, I hereby appeal to the Plumbing Board of Appeals the determination made by the Code Official relative to the interpretation of _____, which provides that _____.

The Appellant proposes _____
_____
_____
_____
_____
_____
_____

In order that I might perform the Plumbing work for the above project as proposed and shown on the attachments.

Appellant is advised to prepare the case and submit any documentation in support of the appeal. Appellant and/or an authorized agent may appear to present reasons for granting an appeal at the time of the scheduled meeting.

_____
Signature of owner or appellant   Date

Fee $_____ Paid _____ Rec't# _____   Meeting Date _____

Plumbing Permit # _____   Date _____

Building Permit # _____   Case # P_____

The jurisdiction must create an atmosphere of trust in which the contractors feel comfortable appealing those decisions of code officials that they truly believe are contrary to the intent of the code.

Topic: Membership of Appeals Board   **Category:** Administration

Reference: IPC 109.2   **Subject:** Means of Appeal

Code Text: *The board of appeals shall consist of five members appointed by the chief appointing authority as follows: one for 5 years, one for 4 years, one for 3 years, one for 2 years and one for 1 year. Thereafter, each new member shall serve for 5 years or until a successor has been appointed.*

Discussion and Commentary: The board of appeals is to consist of five members appointed by the *chief appointing authority*—typically, the mayor or city manager. One member is to be appointed for five years, one for four, one for three, one for two and one for one year. This method of appointment allows for a smooth transition of board of appeals members, allowing continuity of action over the years.

Because of their small size, many jurisdictions combine all of their boards into one board of appeals, which can hear the appeals on various codes such as the plumbing code, mechanical code, electrical code and building code.

Code Text: *The board shall modify or reverse the decision of the code official by a concurring vote of three members.*

Discussion and Commentary: A concurring vote of three members of the board is needed to modify or reverse the decision of the code official. This is obviously a simple majority when the board consists of five members and all five are present at the meeting. If the jurisdiction modifies the composition and number of board members, the number *three* must be modified accordingly.

The jurisdiction must also establish rules as to what constitutes a quorum. Each meeting of the board of appeals must have a quorum to be able to legally conduct business.

Quiz

Study Session 2
IPC Sections 106 – 109

1. The code official shall require construction documents and specifications to be prepared by a registered design professional when_____.

    a. the proposed construction does not comply with the code

    b. the proposed construction is in a flood hazard area

    c. required by state law

    d. special conditions exist

    Reference _____

2. Which of the following information is not normally identified on a permit application?

    a. an alternative engineered design

    b. floor plans and riser diagrams

    c. the proposed occupancy

    d. description of the work

    Reference _____

3. Who is responsible for inspection scheduling?

    a. the permit holder   b. the code official

    c. the utility authority   d. the permit clerk

Reference _____

4. Upon inspection of work and a finding of noncompliance with minimum code provisions, the code official would issue which of the following?

    a. stop work order   b. condemnation order

    c. notice of violation order   d. public nuisance order

Reference _____

5. Which of the following jobs is exempt from requirements for permits?

    a. replacement of a water heater

    b. installation of a cleanout on a building sewer

    c. replacement of concealed water piping

    d. removal and re-installation of a water closet

Reference _____

6. When an ordinance of the government jurisdiction specifies requirements for work that is exempt from permits in the code, how is the conflict resolved?

    a. The code exemption prevails.

    b. The local ordinance prevails.

    c. An appeals board must resolve the conflict.

    d. The exemption prevails if authorized by the code official.

Reference _____

7. A Board of Appeals has the authority to modify or reverse a decision of the code official by a _____.

    a. simple majority vote

    b. unanimous vote

    c. petition to an administrative officer

    d. order for a court review

Reference _____

8. Permits expire if work is not commenced within _____ days of issuance.

   a. 30   b. 60

   c. 120   d. 180

   Reference _____

9. An application for an appeal of the code official's decision must be filed within how many days after receipt of notice of the decision?

   a. 10   b. 20

   c. 30   d. 60

   Reference _____

10. A violation that continues for ten days after notice has been served would be considered _____ violation(s).

    a. one   b. two

    c. five   d. ten

    Reference _____

11. Which of the following tasks would not require testing of the plumbing system as a part of an inspection?

    a. repair of an underground water service

    b. replacement of a section of water service pipe

    c. replacement of an individual vent pipe

    d. alteration of a water supply branch pipe

    Reference _____

12. Which of the following is grounds for suspension or revocation of a permit?

    a. incomplete construction documents

    b. an inspection failure

    c. expiration

    d. a falsified application

    Reference _____

13. When are periodic inspections required?

    a. when the inspector has reason to doubt the integrity of an installation

    b. when an approved alternative engineered design is authorized

    c. when a temporary plumbing system is set up for exhibition purposes

    d. when a rough-in inspection has revealed a failure due to leaks

Reference _____

14. The duly appointed code official does not have the authority to _____.

    a. interpret provisions of the code

    b. enter a building or premises without permission

    c. grant extensions for permits that have exceeded 180 days

    d. rule on an application for appeal of a code interpretation

Reference _____

15. Which of the following is the code official required to prevent?

    a. use of an unsafe plumbing system

    b. use of materials not specifically prescribed by the code

    c. re-use of tested and reconditioned equipment and devices

    d. plumbing systems that are not specifically prescribed by the code

Reference _____

16. A plumbing system design that is not specifically regulated by the plumbing code, but performs in accordance with the intent of the code, is a _____.

    a. code official selected design

    b. alternative engineered design

    c. nonconforming design

    d. nonregulated design

Reference _____

17. A procedure for approval of a material used in a plumbing system, whereby an analysis is performed and documentation is received from an approved laboratory, that reports conformance with specified requirements, is _____.

    a. third-party certification

    b. third-party classification

    c. third-party approval.

    d. third-party testing

Reference _____

18. When a plumbing permit is issued, the construction documents shall be stamped _____.

    a. APPROVED FOR CONSTRUCTION

    b. APPROVED

    c. ACCEPTED AS REVIEWED

    d. REVIEWED FOR CODE COMPLIANCE

Reference _____

19. One set of construction documents shall be kept _____ until completion of the project.

    a. at the job site

    b. by the permit applicant

    c. by the plumbing contractor

    d. by the design professional

Reference _____

20. The board of appeals is not authorized to rule on an appeal based on a claim that _____.

    a. the provisions of the code do not fully apply

    b. a code requirement should be waived

    c. the rules have been incorrectly interpreted

    d. a better form of construction is proposed

Reference _____

21. Plumbing systems shall be tested, with tests made by the permit holder and observed by the _____.

    a. design professional

    b. plumbing contractor

    c. code official

    d. plumbing inspector

Reference _____

22. Construction documents for buildings more than _____ stories in height shall indicate the materials and methods for maintaining the required fire-resistance rating.

    a. 2   b. 3

    c. 4   d. 6

Reference _____

23. The code official may allow temporary connection to the utility source _____.

    a. pending the outcome of an appeal

    b. for occupancy

    c. for construction purposes

    d. for testing plumbing systems

Reference _____

24. The code official has authority to revoke a permit when _____.

    a. the permit was issued in error

    b. required inspections have not been performed

    c. the construction documents are in violation of the code

    d. inaccurate information was provided at the time of issuance

Reference _____

25. When a stop work order is issued, it shall be given to any of the following individuals except the _____.

    a. owner

    b. owner's agent

    c. permit applicant

    d. person doing the work

Reference _____

2006 IPC Chapter 3
General Regulations

OBJECTIVE: To develop an understanding of those general provisions regarding plumbing systems that are not specifically addressed in other chapters of the code. To develop an understanding of the provisions that apply to materials in all plumbing applications.

REFERENCE: Chapter 3, 2006 *International Plumbing Code*

KEY POINTS:
- Which types of plumbing installations are regulated by this chapter?
- Which items are required to be connected to the drainage system?
- Which items are required to be connected to the water supply?
- Which locations cannot be used for the location of plumbing systems?
- When is it permitted to introduce industrial wastes into a public sewer?
- Which standards apply to materials, and how are the materials to be installed?
- Are all plumbing products and materials required to comply with the code?
- What governs when plumbing products and materials should be tested or certified?
- How should openings in walls, floors and ceilings be treated to prevent the entry of rodents?
- When sleeves are used to protect pipes, what additional areas or items need to be addressed?
- Which areas are prohibited from having a pipe installed in them without protection?
- When are shield plates required?
- Under what circumstances is backfill required?
- What material is to be used for backfill, and how is the material placed?
- Is the pipe permitted to rest on rock at any point?
- When over excavation is required, what is the minimum depth?
- When a concrete foundation is used, what type of bedding is required to provide uniform load-bearing support?
- What type of loading may cause damage to the pipe?
- What two problems may occur when earth tunnels are used?
- Which areas of a building are identified by this chapter for alteration during the installation of a plumbing system?

KEY POINTS: • What regulates framing members being cut, notched or bored?
(Cont'd) • What rating is required for penetrations of rated assemblies?
- If a truss is modified, what must occur?
- What piping is required to be supported?
- Which types of materials are permitted to be used for hangers, anchors and supports?
- Which table provides the requirements for the spacing of both horizontal and vertical hangers?
- What defines the design flood elevation? (Definition: The elevation of the *design flood*, including wave height, relative to the datum specified on the community's legally designated flood hazard map).
- Which types of loads and stresses are considered in flood zones?
- Are washrooms and toilet rooms required to be illuminated and ventilated?
- Are there elements of buildings where piping, fixtures or equipment cannot be located?
- Which code regulates the interior surfaces of toilet rooms?
- Are toilet facilities required for construction workers?
- What two methods may be used to test a water supply system?

Topic: Scope

Reference: IPC 301.1

Category: General Regulations

Subject: General Provisions

Code Text: *The provisions of Chapter 3 shall govern the general regulations regarding the installation of plumbing not specific to other chapters.*

Discussion and Commentary: Unlike other chapters of the *International Plumbing Code*, the requirements included in Chapter 3 are not necessarily interrelated. Many regulations are not specific plumbing requirements, but relate to the overall plumbing system. The provisions covered in this chapter apply to most plumbing systems and various parts of the plumbing system.

One such issue not directly related to the performance of plumbing and covered in Chapter 3 is *rodentproofing*. This section deals with the overall health of the building occupants and does not directly impact the performance of the plumbing system.

Topic: System Installation   **Category:** General Regulations

Reference: IPC 301.2   **Subject:** General Provisions

Code Text: *Plumbing shall be installed with due regard to preservation of the strength of structural members and prevention of damage to walls and other surfaces through fixture usage.*

Discussion and Commentary: The installation of plumbing fixtures should not result in a reduction of the structural integrity of building components because of cutting, notching or boring. Installers and code officials alike should inspect fixtures and piping to verify that the installation and use of these components do not have adverse effects on the structure.

Building structural systems are either engineered or constructed according to the conventional construction provisions of the building code. In either case, excessive cutting, notching, boring or otherwise altering the original structural components could negatively impact the performance of the structural system.

Code Text: *Plumbing systems shall not be located in an elevator shaft or in an elevator equipment room.*

Exception: Floor drains, sumps and sump pumps shall be permitted at the base of the shaft provided they are indirectly connected to the plumbing system.

Discussion and Commentary: Plumbing components within the elevator pit, such as floor drains, sumps and sump pumps indirectly connected to the drainage system, are allowed within the elevator shafts. An indirect connection is required to prevent sewage from backing up into the elevator shaft. Other plumbing systems are strictly prohibited in these spaces on account of inaccessibility for repairs and the possible water damage that could be caused to the elevator equipment if a leak developed in the piping system.

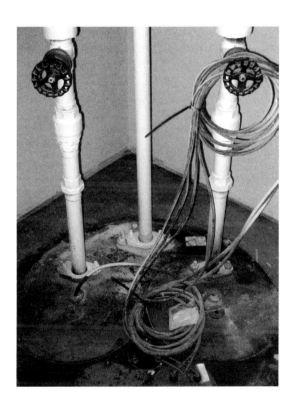

Shafts are in general considered protected elements, and as such the building code has strict limitations on penetrations by pipes, ducts or other building components except in very limited cases or in cases where such penetrations are needed for the service or performance of the shaft itself.

Code Text: *All plastic pipe, fittings and components shall be third-party certified as conforming to NSF 14.*

Discussion and Commentary: All plastic piping, fittings and plastic pipe-related components, including solvent cements, primers, tapes, lubricants and seals used in plumbing systems, are required to be tested and certified as conforming to NSF 14. This includes all water service, water distribution, drainage piping and fittings, and plastic piping system components, including but not limited to pipes, fittings, valves, joining materials, gaskets and appurtenances. This section does not apply to components that only include plastic parts such as brass valves with a plastic stem, or to fixture fittings such as fixture stop valves. Plastic piping systems, fittings and related components intended for use in the potable water supply system must comply with NSF 61 in addition to NSF 14.

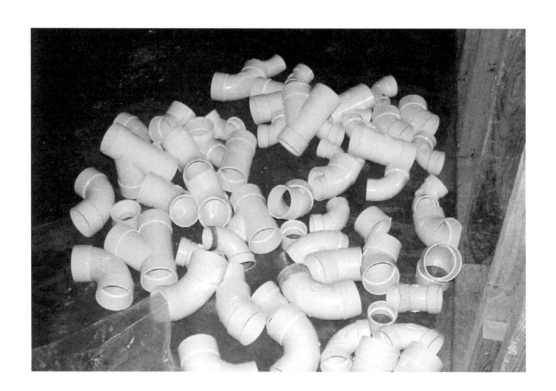

NSF *International Standards* are developed under the American National Standards Institute (ANSI) Consensus process. This process allows participation by all interested or affected segments of the industry, and presents a balanced representation for committee decisions.

Topic: Third-party Testing and Certification

Category: General Regulations

Reference: IPC 303.4

Subject: Materials

Code Text: *All plumbing products and materials shall comply with the referenced standards, specifications and performance criteria of the* International Plumbing Code *and shall be identified in accordance with Section 303.1. When required by Table 303.4, plumbing products and materials shall either be tested by an approved third-party testing agency or certified by an approved third-party certification agency.*

Discussion and Commentary: This section requires that all plumbing products and materials comply with the referenced standards. However, the provisions contained in Section 105.2 regarding the evaluation and approval of alternative materials, methods and equipment are still applicable. The code includes specific requirements for plumbing products in regard to third-party certification and testing. Table 303.4 lists these requirements. *Third-party certified* indicates that the minimum level of quality required by the appropriate standard is maintained, and the product is often referred to as *listed*. *Third-party tested* indicates a product that has been tested by an approved testing laboratory and found to be in compliance with the appropriate standard.

TABLE 303.4
PRODUCTS AND MATERIALS REQUIRING THIRD-PARTY TESTING AND THIRD-PARTY CERTIFICATION

PRODUCT OR MATERIAL	THIRD-PARTY CERTIFIED	THIRD-PARTY TESTED
Potable water supply system components and potable water fixture fittings	Required	—
Sanitary drainage and vent system components	Plastic pipe, fittings and pipe-related components	All others
Waste fixture fittings	Plastic pipe, fittings and pipe-related components	All others
Storm drainage system components	Plastic pipe, fittings and pipe-related components	All others
Plumbing fixtures	—	Required
Plumbing appliances	Required	—
Backflow prevention devices	Required	—
Water distribution system safety devices	Required	—
Special waste system components	—	Required
Subsoil drainage system components	—	Required

Third parties providing certification must be approved (acceptable to the code official). One such method for approval is accreditation by the International Accreditation Service (IAS).

Code Text: *Pipes passing through or under walls shall be protected from breakage.*

Discussion and Commentary: When piping is installed through or under a wall assembly, it is subjected to any loading conditions that the wall is resisting; therefore, protection from the effects of building structural loads must be considered. This protection can be achieved through the use of a sleeve or a relieving arch.

Plumbing pipes are designed to carry only the load of the contents passing through them and no other loads. The code provides for protection of pipes in various situations such as support under pipes, hanger spacing and sleeves through walls.

Code Text: *Annular spaces between sleeves and pipes shall be filled or tightly caulked in an approved manner. Annular spaces between sleeves and pipes in fire-resistance-rated assemblies shall be filled or tightly caulked in accordance with the* International Building Code.

Discussion and Commentary: The annular space created between a sleeve and a pipe in a nonfire-resistance-rated assembly must be filled with a material that prevents structural loading of the pipe and rodent infiltration. This material is typically caulk with coal tar or an asphaltic compound with some degree of flexibility. Materials used to fill the annular space are to be compatible with the pipe as well as the sleeve material.

Annular spaces created by penetrations in fire-resistance-rated assemblies must also be filled and must maintain the fire-resistance rating of the assembly. The fire-resistance rating can be protected by providing a through-penetration system, which can consist of caulks, intumescent materials and sleeves installed around the penetrating pipe. All through-penetration systems must be tested in accordance with ASTM E 814 as provided in the *International Building Code* (IBC).

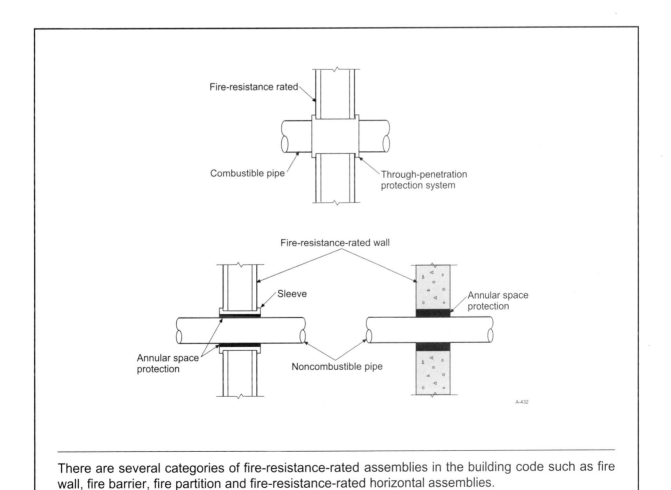

There are several categories of fire-resistance-rated assemblies in the building code such as fire wall, fire barrier, fire partition and fire-resistance-rated horizontal assemblies.

Code Text: *Any pipe that passes under a footing or through a foundation wall shall be provided with a relieving arch, or a pipe sleeve pipe shall be built into the foundation wall. The sleeve shall be two pipe sizes greater than the pipe passing through the wall.*

Discussion and Commentary: Piping installed within or under a footing or foundation wall must be structurally protected from any transferred loading from the footing or foundation wall. This protection may be provided through the use of a relieving arch or a pipe sleeve.

When a sleeve is used, it should be sized such that it is two pipe sizes larger than the penetrating pipe. For example, a 4-inch (102 mm) penetrating pipe would require a 6-inch (152 mm) sleeve. This space will allow for any differential movement of the pipe. By providing structural protection to the piping system, the piping will not be subjected to undue stresses that could cause it to rupture and leak.

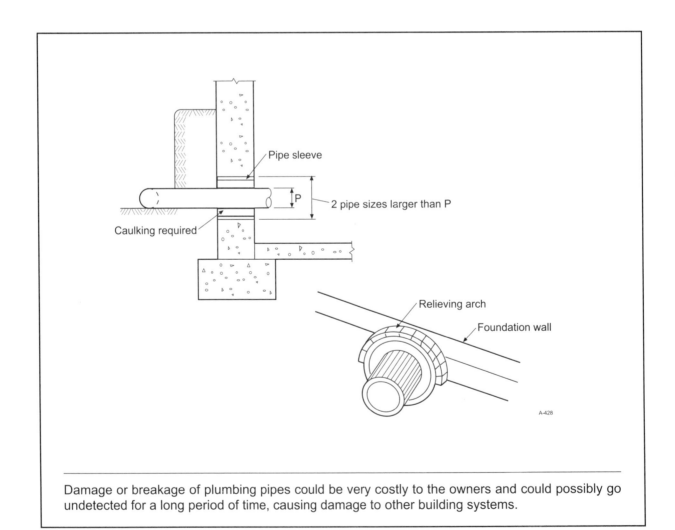

Damage or breakage of plumbing pipes could be very costly to the owners and could possibly go undetected for a long period of time, causing damage to other building systems.

Topic: Freezing

Reference: IPC 305.6

Category: General Regulations

Subject: Protection of Pipes and Components

Code Text: *Water, soil and waste pipes shall not be installed outside of a building, in attics or crawl spaces, concealed in outside walls, or in any other place subjected to freezing temperature unless adequate provision is made to protect such pipes from freezing by insulation or heat or both. Exterior water supply system piping shall be installed not less than 6 inches (152 mm) below the frost line and not less than 12 inches (305 mm) below grade.*

Discussion and Commentary: When a water or drain pipe is installed in an exterior wall or unheated space, such as a crawl space or attic, freeze protection can be achieved through thermal insulation. No amount of insulation alone (without a heat source) can prevent freezing; insulation can only delay freezing by slowing the rate of heat loss.

Hose bibbs and wall hydrants located on the exterior wall must be protected when installed in areas subject to freezing temperature. This can be accomplished by installing devices, such as freezeproof hose bibbs, that locate the valve seat within the heated space and allow residual water within the hydrant to drain after the valve is closed.

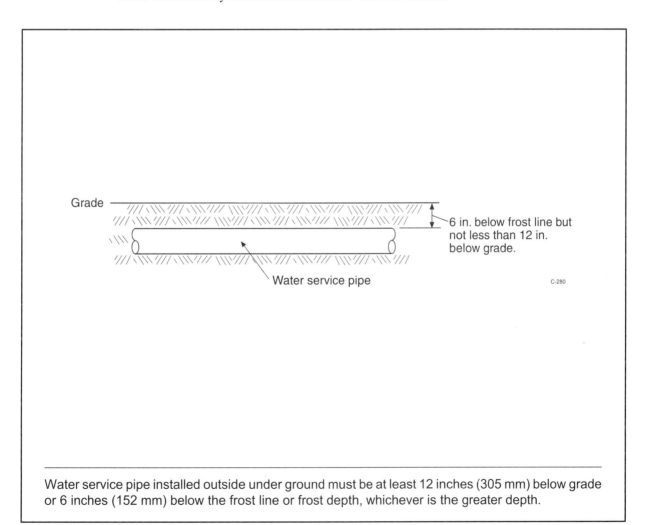

Grade

6 in. below frost line but not less than 12 in. below grade.

Water service pipe

C-280

Water service pipe installed outside under ground must be at least 12 inches (305 mm) below grade or 6 inches (152 mm) below the frost line or frost depth, whichever is the greater depth.

Code Text: *Joints at the roof and around vent pipes, shall be made water tight by the use of lead, copper, galvanized steel, aluminum, plastic or other approved flashings or flashing material. Exterior wall openings shall be made water tight.*

Discussion and Commentary: Where a pipe penetrates a roof or an exterior wall, the annular space around the pipe must be sealed, or the pipe must be provided with flashing to prevent the entry of moisture. If this annular space is improperly sealed or not sealed at all, moisture from precipitation can enter the structure and damage the surrounding structure and finishes.

Because exterior walls are typically not subjected to the same severity of exposure as a roof, the space around pipes that penetrate exterior walls is not required to be sealed with flashing; rather, it is only required to be made water tight. Openings in exterior walls, therefore, can be sealed with an approved sealant, mechanical device, flashing or other approved method that will prevent penetration into the structure.

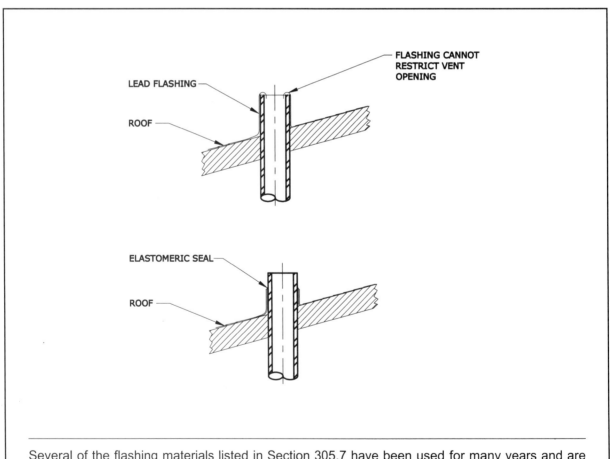

Several of the flashing materials listed in Section 305.7 have been used for many years and are recognized as effective moisture penetration protection. This section also provides for the use of new and innovative materials that are not specifically listed when approved by the code official.

Topic: Trench Location

Category: General Regulations

Reference: IPC 307.5

Subject: Structural Safety

Code Text: *Trenches installed parallel to footings shall not extend below the 45-degree (0.79 rad) bearing plane of the footing or wall.*

Discussion and Commentary: A footing requires a minimum load-bearing area to distribute the weight of the building. This load-bearing distribution plane extends downward at approximately a 45-degree (0.79 rad) angle from the base of the footing. Water and sewer piping must not be installed below this load-bearing plane, nor can excavation for the installation of pipe extend below the plane, so as not to affect the load capacity of the footing, or cause the excavation to collapse.

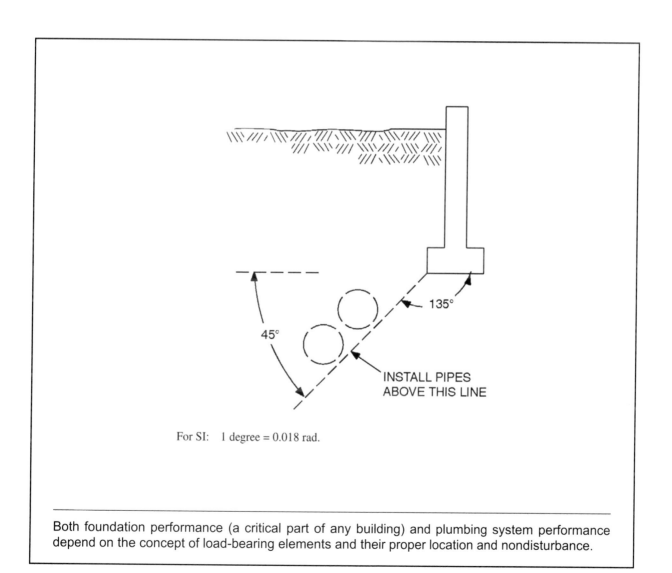

For SI: 1 degree = 0.018 rad.

Both foundation performance (a critical part of any building) and plumbing system performance depend on the concept of load-bearing elements and their proper location and nondisturbance.

Code Text: *Anchorage shall be provided to restrain drainage piping from axial movement.*

Discussion and Commentary: This section requires a method of resisting axial movement of piping systems in order to prevent joint separation, regardless of the type of fittings or connections used. In particular, mechanical couplings using an elastomeric seal (typically hubless piping systems) have a limited ability to resist axial movement (pulling apart); therefore, pipe restraints must be provided to prevent joint separation. The required anchorage locations of axial restraints are specified in Section 308.7.1. Such joints also have a limited ability to resist shear forces. The hanger and support system must, therefore, prevent the couplings and connections from being subjected to shear forces that can damage the joint.

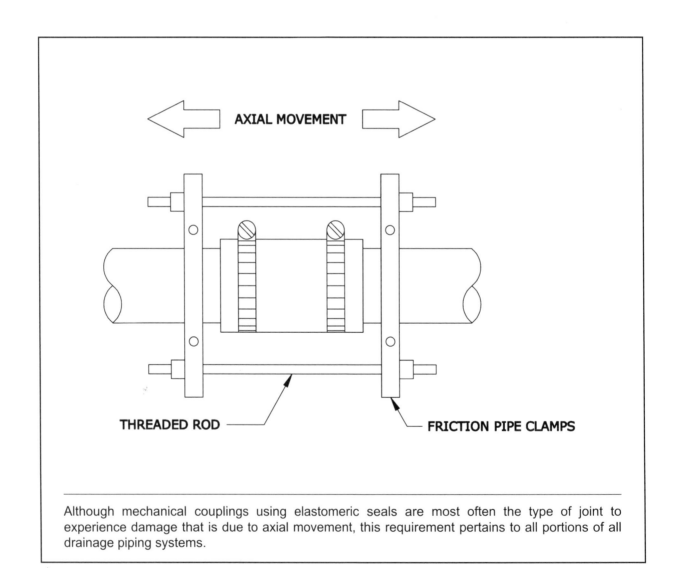

Although mechanical couplings using elastomeric seals are most often the type of joint to experience damage that is due to axial movement, this requirement pertains to all portions of all drainage piping systems.

Code Text: *Piping, fixtures or equipment shall not be located in such a manner as to interfere with the normal operation of windows, doors or other means of egress openings.*

Discussion and Commentary: Fixtures and piping located in washrooms or toilet rooms cannot adversely affect the operation of building components such as windows and doors. The location of the fixtures and piping should not interfere with or block an occupant's egress to or from that space. Obstructions caused by fixtures or piping can cause injury to the occupants or seriously hamper egress in a fire condition.

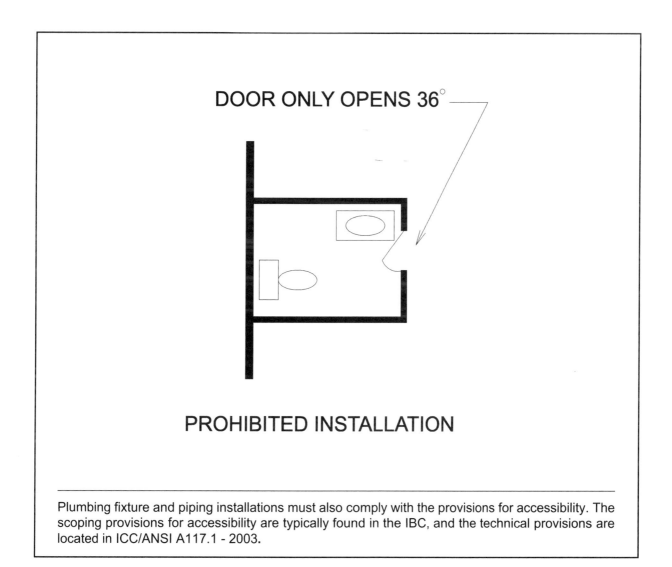

Plumbing fixture and piping installations must also comply with the provisions for accessibility. The scoping provisions for accessibility are typically found in the IBC, and the technical provisions are located in ICC/ANSI A117.1 - 2003.

Code Text: *Each water closet utilized by the public or employees shall occupy a separate compartment with walls or partitions and a door enclosing the fixtures to ensure privacy.*

Exceptions:
1. *Water closet compartments shall not be required in a single-occupant toilet room with a lockable door.*
2. *Toilet rooms located in day care and child-care facilities and containing two or more water closets shall be permitted to have one water closet without an enclosing compartment.*

Discussion and Commentary: Lack of privacy places a burden on an individual's physical ability to use bathroom facilities on account of uneasiness, inhibition or indignation. A partitioned compartment can provide the necessary privacy. Also, a single-occupant toilet room with a lockable door can be deemed to provide the required privacy as stated in the first exception. The second exception, allowing an unenclosed water closet in a child-care facility, recognizes the need for people to be able to assist children.

All areas of a toilet room, including toilet partitions, must be designed such that the toilet room complies with the accessibility requirements.

| **Topic:** Urinal Partitions | **Category:** General Regulations |
| **Reference:** IPC 310.5 | **Subject:** Washroom and Toilet Room Requirements |

Code Text: *Each urinal utilized by the public or employees shall occupy a separate area with walls or partitions to provide privacy. The construction of such walls or partitions shall incorporate waterproof, smooth, readily cleanable and nonabsorbent finish surfaces. The walls or partitions shall begin at a height not more than 12 inches (305 mm) from and extend not less than 60 inches (1524 mm) above the finished floor surface. The walls or partitions shall extend from the wall surface at each side of the urinal a minimum of 18 inches (457 mm) or to a point not less than 6 inches (152 mm) beyond the outermost front lip of the urinal measured from the finished back wall surface, whichever is greater.*

Exceptions:
1. *Urinal partitions shall not be required in a single occupant or unisex toilet room with a lockable door.*
2. *Toilet rooms located in day care and child care facilities and containing two or more urinals shall be permitted to have one urinal without partitions.*

Discussion and Commentary: Privacy concerns for toilet facilities are not limited to water closets. Urinals will need to have privacy partitions installed as well.

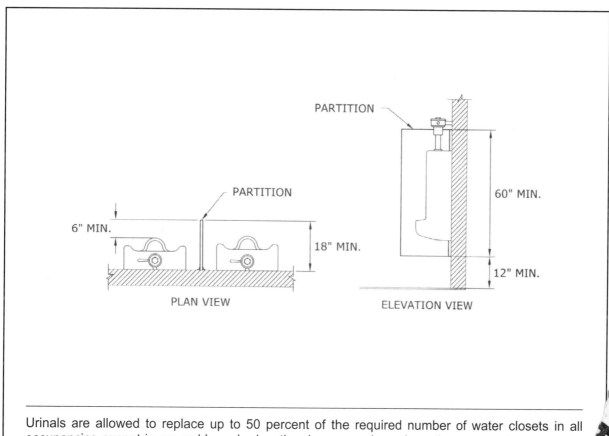

Urinals are allowed to replace up to 50 percent of the required number of water closets in all occupancies except in assembly and educational occupancies, where they can replace up to 67 percent of the required number of water closets.

Code Text: *Upon completion of a section of or the entire water supply system, the system, or portion completed shall be tested and proved tight under a water pressure not less than the working pressure of the system; or, for piping systems other than plastic, by an air test of not less than 50 psi (344 kPa). This pressure shall be held for at least 15 minutes. The water utilized for tests shall be obtained from a potable source of supply. The required tests shall be performed in accordance with this section and Section 107.*

Discussion and Commentary: Plastic piping manufacturers and engineers warn against testing with air because air is compressible and therefore stores energy. During the test with air, it is possible to have a loose fitting separate. The stored air could then cause the separated piping or fitting to become a projectile.

There have been reports of injuries and deaths attributed to testing with high-pressure air. Caution should be exercised if air is to be used for testing.

Quiz

Study Session 3
IPC Chapter 3

1. Which of the following is required of all fixtures that discharge wastewater?

    a. Fixtures must be connected directly or indirectly to the sanitary drainage system.

    b. Fixtures must discharge by a direct connection to the sanitary drainage system.

    c. Fixtures must discharge to an approved public sewer system.

    d. Fixtures must first discharge to a pretreatment system.

    Reference _____

2. When there is a conflict between the requirements of the code and the conditions of a listing or the manufacturer's installation instructions, which is the prevailing minimum requirement?

    a. the IPC

    b. the listing requirements

    c. the most restrictive requirement

    d. the least restrictive requirement

    Reference _____

3. Which of the following is allowed at the base of an elevator shaft?

    a. a reduced pressure principle backflow preventer

    b. a hose bibb that is protected by a vacuum breaker

    c. a floor drain, when directly connected to a storm drain

    d. a floor drain, when indirectly connected to the plumbing system

    Reference _____

4. Which of the following are required to be third-party certified?

    a. cast iron pipe

    b. plumbing fixtures

    c. potable water supply components

    d. special waste system components

    Reference _____

5. Which of the following is not approved for protection against rodents entering a structure?

    a. pipe escutcheons   b. strainer plates

    c. building traps   d. caulking

    Reference _____

6. The minimum burial depth of a water service pipe is _____.

    a. 12 inches below the frost line

    b. 12 inches below grade

    c. 8 inches below the frost line

    d. 8 inches below grade

    Reference _____

7. Which of the following is prohibited for use to support and maintain grade of drainage piping in a trench?

    a. gravel backfill   b. sand backfill

    c. clay soil   d. blocking

Reference _____

8. When a 2-inch pipe is installed through a bored hole in a nonbearing 2 x 4 stud wall, which of the following is required?

    a. sleeving   b. shield plates

    c. protective wrapping   d. anchors at the penetrations

Reference _____

9. What is the maximum horizontal hanger spacing for $1^1/_4$-inch polypropylene pipe and tubing?

    a. 32 inches   b. 48 inches

    c. 8 feet   d. 10 feet

Reference _____

10. For water piping adjacent and parallel to a wall footing, and installed at a depth of 12 inches below the bottom of the footing, the water piping shall be located not less than _____ inches from the footing.

    a. 24   b. 12

    c. 6   d. 18

Reference _____

11. Which of the following codes regulate the installation of piping exposed in plenums?

    a. *International Fire Code* (IFC)

    b. *International Fuel Gas Code* (IFGC)

    c. *International Mechanical Code* (IMC)

    d. *International Plumbing Code* (IPC)

Reference _____

12. A 4-inch polypropylene water pipe, with compression sleeve joints, suspended from a ceiling around a room, and parallel to the walls within the room, is required to be supported by which of the following:

    a. hangers only

    b. hangers and sway bracing only

    c. hangers, sway bracing, and coupling restraints only

    d. hangers, sway bracing, coupling restraints and thrust blocks

    Reference _____

13. The maximum horizontal spacing of support hangers for 10-foot lengths of cast-iron soil pipe suspended from a ceiling is _____.

    a. 5 feet

    b. 10 feet

    c. at every change in direction

    d. at every change in direction greater than 45 degrees

    Reference _____

14. Privacy partitions for urinals are not required in a _____.

    a. employee restroom

    b. child-care facility restroom

    c. lockable unisex toilet room

    d. toilet room for females

    Reference _____

15. Which of the following testing criteria is not approved for a PVC water line, supplied with a pressure of 60 PSI from the supply main?

    a. a water pressure test of 80 PSI

    b. an air pressure test of not less than 60 PSI

    c. a partial test on a completed and isolated section of the piping system

    d. a pressure test using water supplied from a portable water supply storage tank

    Reference _____

16. All of the following conditions would cause a backflow prevention assembly to have a required inspection and test, except _____.

    a. when newly installed

    b. the anniversary date of the installation

    c. a new connection of a potential source of contamination, downstream from the assembly

    d. the relocation of an existing assembly that was tested previously, 6 months prior to relocation

Reference _____

17. The minimum required grade for a condensate pipe from a heating appliance is _____.

    a. 1 percent   b. 2 percent

    c. $^1/_4$ inch per foot   d. $^1/_2$ inch per foot

Reference _____

18. Which of the following is required on a central heating and A© unit that does not have a secondary drain or means to install an auxiliary drain pan?

    a. a water level site glass

    b. an external secondary drain line bypass

    c. an external water level monitoring device, in the drain line

    d. a water level monitoring device inside the primary drain pan

Reference _____

19. Structures located in flood hazard areas are required to have which of the following installed above the design flood elevation?

    a. anchored drain pipes

    b. tank-type water closet

    c. sealed manhole covers

    d. sealed covers on potable water well casings

Reference _____

20. Which of the following is required on all construction sites?

    a. water dispensers   b. emergency showers

    c. emergency eyewashes   d. toilet facilities

Reference _____

21. Under what circumstance is the boring of a hole in an engineered truss allowed?

    a. when approved by the code official

    b. when approved by a registered design professional

    c. when no additional load is being exerted on the truss

    d. when the hole does not exceed 40 percent of the component

Reference _____

22. What is the minimum time interval for a pressure test on the water supply system?

    a. 15 minutes

    b. 30 minutes

    c. 1 hour

    d. the time necessary to determine if there are leaks

Reference _____

23. A procedure for approval of a material used in a plumbing system, whereby an analysis is performed and documentation is received from an approved laboratory that reports conformance with specified requirements, is _____.

    a. third-party certification   b. third-party classification

    c. third-party approval   d. third-party testing

Reference _____

24. Where water supply systems other than plastic are tested with air, the test pressure shall be not less than _____ psi.

    a. 10   b. 20

    c. 30   d. 50

Reference _____

25. Where rock is encountered in trenching, it shall be removed to a minimum of _____ inches below the installation level of the bottom of the pipe.

a. 2

b. 3

c. 4

d. 6

Reference _____

Study Session

2006 IPC Sections 401 – 411
Fixtures, Faucets and Fixture Fittings I

OBJECTIVE: To develop an understanding of how the code describes the number and types of plumbing fixtures needed for building occupants and fixture protection requirements.

REFERENCE: Sections 401 through 411, 2006 *International Plumbing Code*

KEY POINTS:
- What is used to determine the number and types of fixtures in a structure?
- Does the code regulate maximum flow rates and flush volume of fixtures?
- In addition to the minimum weight, what standard applies to sheet copper?
- How is occupancy classification determined?
- Are urinals permitted to be substituted for water closets?
- What is the maximum travel distance for an employee to a toilet facility?
- What are the requirements for the locations of the toilet facilities?
- Proper installation of the water supply lines is required to prevent what from occurring?
- Closet screws and bolts are required to be of what materials?
- What type of carrier is required to support a wall-hung water closet bowl?
- When an overflow is provided for a fixture, how is the waste designed?
- Where may slip joints be installed?
- Does the code permit the waste from a clothes washer to discharge into a laundry sink?
- What standards apply to various plumbing fixtures?
- What type of protection is required for the water supply to a bidet?
- What type of protection is required for the water supply to a dishwasher?
- What are the standards for drinking fountains and water coolers?

Code Text: *Plumbing fixtures shall be provided for the type of occupancy and in the minimum number shown in Table 403.1. Types of occupancies not shown in Table 403.1 shall be considered individually by the code official. The number of occupants shall be determined by the* International Building Code. *Occupancy classification shall be determined in accordance with the* International Building Code.

Discussion and Commentary: Table 403.1 establishes the minimum number of plumbing fixtures required for each building. The occupant load used for calculating the number of fixtures required is typically the same occupant load used for determining means of egress in the building code. This table provides simple straight-line ratios for determining the minimum number of plumbing fixtures. To aid in the use of the table, the type of building category has been listed by occupancy group classification along with a brief description. The occupancy groups are identical to the classification listed in the building code.

TABLE 403.1
MINIMUM NUMBER OF REQUIRED PLUMBING FIXTURES[a]
(See Sections 403.2 and 403.3)

NO.	CLASSIFICATION	OCCUPANCY	DESCRIPTION	WATER CLOSETS (URINALS SEE SECTION 419.2) MALE	FEMALE	LAVATORIES MALE	FEMALE	BATHTUBS/ SHOWERS	DRINKING FOUNTAIN (SEE SECTION 410.1)	OTHER
1	Assembly (see Sections 403.2, 403.4 and 403.4.1)	A-1[d]	Theaters and other buildings for the performing arts and motion pictures	1 per 125	1 per 65	1 per 200		—	1 per 500	1 service sink
		A-2[d]	Nightclubs, bars, taverns, dance halls and buildings for similar purposes	1 per 40	1 per 40	1 per 75		—	1 per 500	1 service sink
			Restaurants, banquet halls and food courts	1 per 75	1 per 75	1 per 200		—	1 per 500	1 service sink
		A-3[d]	Auditoriums without permanent seating, art galleries, exhibition halls, museums, lecture halls, libraries, arcades and gymnasiums	1 per 125	1 per 65	1 per 200		—	1 per 500	1 service sink
			Passenger terminals and transportation facilities	1 per 500	1 per 500	1 per 750		—	1 per 1,000	1 service sink
			Places of worship and other religious services.	1 per 150	1 per 75	1 per 200		—	1 per 1,000	1 service sink

(Continued)

The means of egress occupant load of a building does not always reflect typical day-to-day occupant loads; however, the table takes this into account by modifying the values for determining the number of fixtures.

Code Text: *Where plumbing fixtures are required, separate facilities shall be provided for each sex.*

Exceptions:
1. *Separate facilities shall not be required for dwelling units and sleeping units.*
2. *Separate facilities shall not be required in structures or tenant spaces with a total occupant load, including both employees and customers, of 15 or less.*
3. *Separate facilities shall not be required in mercantile occupancies in which the maximum occupant load is 50 or less.*

Discussion and Commentary: Separate facilities must be provided for each sex—meaning a separate women's room and men's room.

Single-occupant toilet rooms for use by both sexes are permitted for:
1. Private residential uses, such as within dwelling units and hotel guestrooms;
2. Employees in buildings in which 15 or fewer people are employed; and
3. Small establishments in which the occupant load, including customers and employees, is 15 or fewer.

Where the required number of fixtures based on Table 403.1 is one water closet and one lavatory, two of each fixture must be provided. One group of fixtures must be installed in the women's room and one in the men's room.

Topic: Required Public Toilet Facilities   **Category:** Fixtures, Faucets and Fixture Fittings

Reference: IPC 403.4   **Subject:** Minimum Plumbing Facilities

Code Text: *Customers, patrons and visitors shall be provided with public toilet facilities in structures and tenant spaces intended for public utilization. The accessible route to public facilities shall not pass through kitchens, storage rooms, closet or similar spaces. Employees shall be provided with toilet facilities in all occupancies. Employee toilet facilities shall be either separate or combined employee and public toilet facilities.*

Discussion and Commentary: The requirement for where employee facilities must be located have been divided into two categories: other covered malls and covered malls. Employee facilities in occupancies other than covered malls must be located in close proximity to the employees' work area.

ACCESSIBLE ROUTE: A continuous, unobstructed path that complies with Chapter 11 of the *International Building Code* (IBC).

Code Text: *In occupancies other than covered malls, the required public and employee toilet facilities shall be located not more than one story above or below the space required to be provided with toilet facilities, and the path of travel to such facilities shall not exceed a distance of 500 feet (152 m).*

Exception: The location and maximum travel distances to required employee facilities in factory and industrial occupancies are permitted to exceed that required by this section, provided that the location and maximum travel distance are approved.

Discussion and Commentary: What is considered to be an employee's working area is subject to interpretation; however, the code clearly provides two conditions that must be met. The required toilet facilities must be located within a travel distance of 500 feet (152 m), and the employee must not be required to travel beyond the next adjacent story above or below his or her working area. Where public facilities are provided in the employee's working area, such facilities can serve as the required employee facilities. Otherwise, separate employee facilities are required.

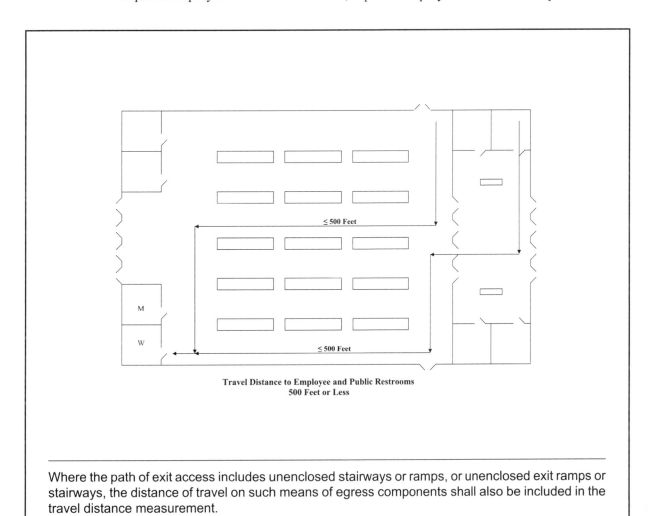

Travel Distance to Employee and Public Restrooms
500 Feet or Less

Where the path of exit access includes unenclosed stairways or ramps, or unenclosed exit ramps or stairways, the distance of travel on such means of egress components shall also be included in the travel distance measurement.

Code Text: *In covered mall buildings, the required public and employee toilet facilities shall be located not more than one story above or below the space required to be provided with toilet facilities, and the path of travel to such facilities shall not exceed a distance of 300 feet (91 440 mm). In covered mall buildings, the required facilities shall be based on total square footage, and facilities shall be installed in each individual store or in a central toilet area located in accordance with this section. The maximum travel distance to central toilet facilities in covered mall buildings shall be measured from the main entrance of any store or tenant space. In covered mall buildings, where employees' toilet facilities are not provided in the individual store, the maximum travel distance shall be measured from the employees' work area of the store or tenant space.*

Discussion and Commentary: In covered malls, the path of travel to required toilet facilities must not exceed a distance of 300 feet (91 440 mm) and is measured from the main entrance of any store or tenant space. This addresses the needs of the elderly, persons with disabilities and children. Additionally, covered malls are frequently very congested, and occupants are unfamiliar with their surroundings. The minimum number of required plumbing facilities is based on total square footage of the mall, including tenant spaces, and must be installed in each individual store or in a central toilet area. This section does not prohibit the installation of separate toilet facilities in individual tenant spaces. However, such facilities are not deductible from the total common facilities required if the facilities in the tenant space are intended for employees only.

COVERED MALL BUILDING: A single building enclosing a number of tenants and occupants, such as retail stores, drinking and dining establishments, entertainment and amusement facilities, passenger transportation terminals, offices and other similar uses wherein two or more tenants have a main entrance into one or more malls. For the purpose of this chapter, anchor buildings shall not be considered a part of the covered mall building.

Topic: Water Supply Protection	**Category:** Fixtures, Faucets and Fixture Fittings
Reference: IPC 405.1	**Subject:** Installation of Fixtures

Code Text: *The supply lines and fittings for every plumbing fixture shall be installed so as to prevent backflow.*

Discussion and Commentary: This section identifies minimal installation requirements for water supply protection. Backflow prevention is perhaps the most important aspect of a plumbing system. All plumbing fixtures, plumbing appliances and water distribution system openings, outlets and connections are potentially capable of contaminating the potable water supply. Backflow can occur as a result of backpressure or backsiphonage, both of which are defined terms in the IPC.

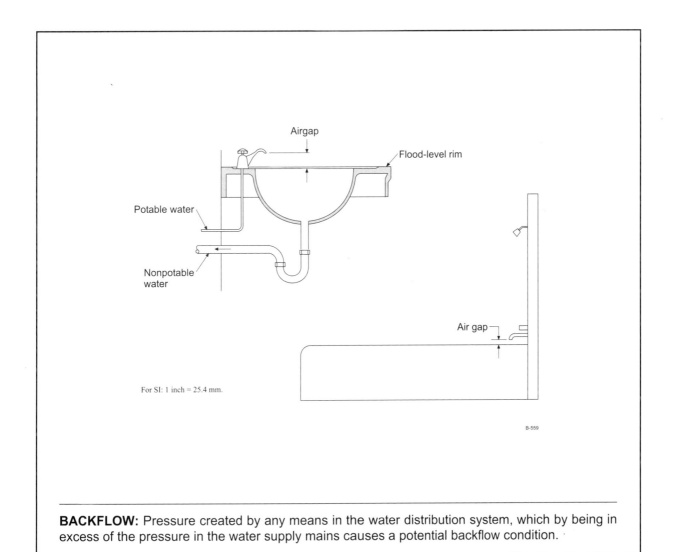

For SI: 1 inch = 25.4 mm.

B-559

BACKFLOW: Pressure created by any means in the water distribution system, which by being in excess of the pressure in the water supply mains causes a potential backflow condition.

Code Text: *A water closet, urinal, lavatory or bidet shall not be set closer than 15 inches (381 mm) from its center to any side wall partition, vanity or other obstruction, or closer than 30 inches (762 mm) center-to-center between adjacent fixtures. There shall be at least a 21-inch (533 mm) clearance in front of the water closet, urinal, lavatory or bidet to any wall, fixture or door. Water closet compartments shall not be less than 30 inches (762 mm) wide and 60 inches (1524 mm) deep.(see Figure 405.3.1)*

Discussion and Commentary: This section establishes minimum clearances for mounting of plumbing fixtures. Adequate clearances are critical for proper and comfortable usability.

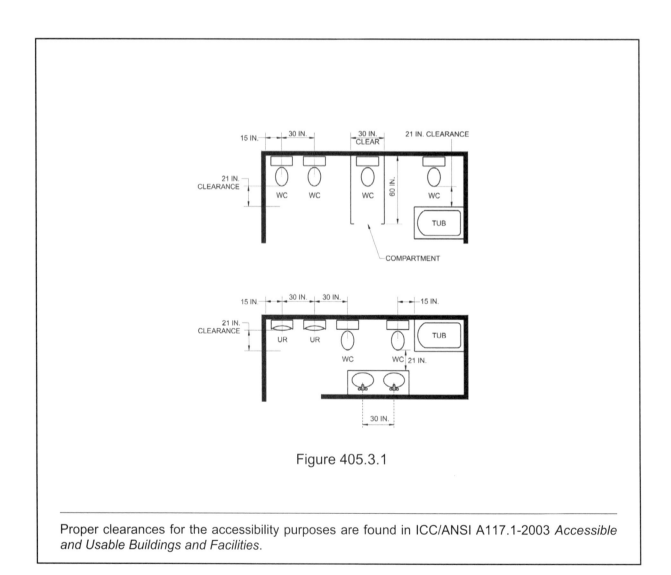

Figure 405.3.1

Proper clearances for the accessibility purposes are found in ICC/ANSI A117.1-2003 *Accessible and Usable Buildings and Facilities.*

Code Text: *Connections between the drain and floor outlet plumbing fixtures shall be made with a floor flange. The flange shall be attached to the drain and anchored to the structure. Connections between the drain and wall-hung water closets shall be made with an approved extension nipple or horn adapter. The water closet shall be bolted to the hanger with corrosion-resistant bolts or screws. Joints shall be sealed with an approved elastomeric gasket, flange-to-fixture connection complying with ASME A112.4.3 or an approved setting compound.*

Discussion and Commentary: The setting compound most commonly used is a wax ring made of beeswax or a synthetic wax. In the past, putty was commonly used. Elastomeric gaskets are also used between fixture outlets (horns) and flanges.

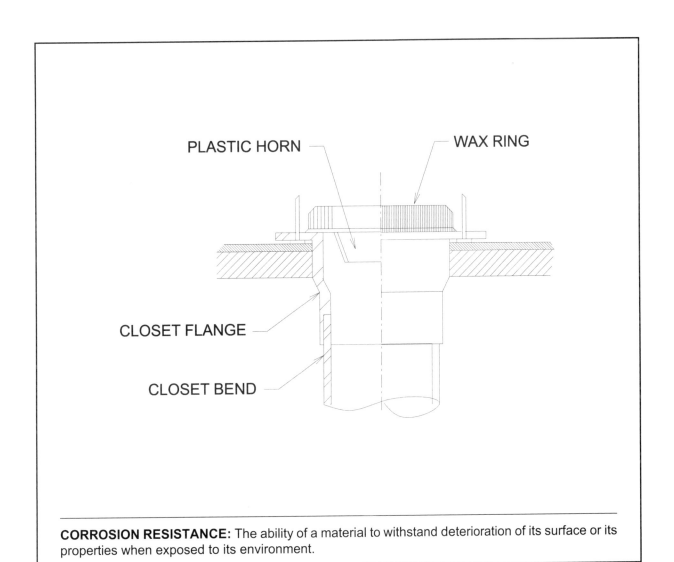

CORROSION RESISTANCE: The ability of a material to withstand deterioration of its surface or its properties when exposed to its environment.

Code Text: *Joints formed where fixtures come in contact with walls or floors shall be sealed.*

Discussion and Commentary: This section addresses the surface connection of the fixture to the floor or wall. The point of contact does not form a joint with any piping material; however, it is still an important area to seal so as to prevent a concealed fouling surface.

The contact joint is sealed with a substance such as plaster of paris, grout or silicon caulking. The exact material selected to seal the joint should be capable of withstanding any anticipated movement of the fixture during its normal use.

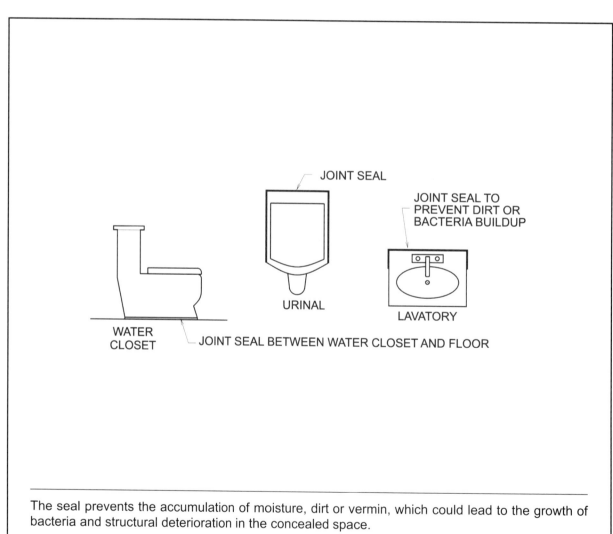

JOINT SEAL

JOINT SEAL TO PREVENT DIRT OR BACTERIA BUILDUP

URINAL

LAVATORY

WATER CLOSET

JOINT SEAL BETWEEN WATER CLOSET AND FLOOR

The seal prevents the accumulation of moisture, dirt or vermin, which could lead to the growth of bacteria and structural deterioration in the concealed space.

Code Text: *The waste from an automatic clothes washer shall discharge through an air break into a standpipe in accordance with Section 802.4 or into a laundry sink. The trap and fixture drain for an automatic clothes washer standpipe shall be a minimum of 2 inches (51 mm) in diameter. The automatic clothes washer fixture drain shall connect to a branch drain or drainage stack a minimum of 3 inches (76 mm) in diameter. Automatic clothes washers that discharge by gravity shall be permitted to drain to a waste receptor or an approved trench drain.*

Discussion and Commentary: A clothes washer must discharge to the drainage system through an air break. A direct connection would allow sewage to back up into the appliance in the event of a blockage in the drain pipe. A clothes washer is usually discharged to an individually trapped standpipe. It is not uncommon, however, to find clothes washer installations where the appliance discharges to a laundry sink or by gravity to a floor drain or trench drain. The actual connection of the discharge pipe needs to be reviewed to verify that the required air break has been approved.

802.4 Standpipes

Standpipes shall be individually trapped. Standpipes shall extend a minimum of 18 inches (457) mm) and a maximum of 42 inches (1066 mm) above the trap weir. Access shall be provided to all standpipes and drains for rodding.

AIR BREAK (Drainage System): A piping arrangement in which a drain from a fixture, appliance or device discharges indirectly into another fixture, receptacle or interceptor at a point below the flood level rim and above the trap seal.

Code Text: *Bathtubs shall have waste outlets a minimum of 1.5 inches (38 mm) in diameter. The waste outlet shall be equipped with an approved stopper.*

Discussion and Commentary: The minimum required diameter of a bathtub waste outlet is 1.5 inches (38 mm). A bathtub is required to have an overflow waste outlet to protect against accidental flooding of the bathroom by an unattended filling operation or to remove displaced water when a bather enters a full tub.

It should be noted that a tub waste and overflow assembly utilizing slip joints must be made accessible in accordance with Section 405.8.

Code Text: *The discharge water temperature from a bidet fitting shall be limited to a maximum temperature of 110°F (43°C) by a water temperature limiting device conforming to ASSE 1070.*

Discussion and Commentary: ASSE 1070, *Performance Requirements for Water-temperature Limiting Devices*, regulates devices intended for this type of application. The device is a thermostatic mixing valve with a maximum temperature limit adjustment.

An air gap or backflow preventer is required for protection of water supply connected to a bidet (Section 408.2).

Code Text: *Drinking fountains shall conform to ASME A112.19.1M, ASME A112.19.2M or ASME A112.19.9M and water coolers shall conform to ARI 1010. Drinking fountains and water coolers shall conform to NSF 61, Section 9. Where water is served in restaurants, drinking fountains shall not be required. In other occupancies, where drinking fountains are required water coolers or bottled water dispensers shall be permitted to be substituted for not more than 50 percent of the required drinking fountains.*

Discussion and Commentary: Drinking fountains must be provided for building occupants and others visiting a building for most occupancies. Water fountains are not required in restaurants if water is available to be served, and in several other occupancies as indicated in Table 403.1. Water fountains must meet minimum standards for safety.

Water coolers and bottled water dispensers can be considered part of the required number of water fountains but cannot account for more than 50 percent of the total number of required water fountains.

Code Text: *Drinking fountains shall not be installed in public restrooms.*

Discussion and Commentary: Drinking fountains are prohibited from being installed in public toilet rooms, so as to prevent the contamination of drinking water. Biological aerosols are present in these rooms and can contaminate exposed surfaces. The drinking fountain must be protected against surface contamination.

CONTAMINATION: An impairment of the quality of the potable water that creates an actual hazard to the public health through poisoning or through the spread of disease by sewage, industrial fluids or waste.

Code Text: *Emergency showers and eyewash stations shall conform to ISEA Z358.1.*

Discussion and Commentary: Water and waste connection requirements are provided in this section for specialized plumbing fixtures, such as emergency showers and eyewash stations.

Emergency showers and eyewash stations are provided in areas where an individual may come in contact with substances that are immediately harmful to the body. These emergency devices provide an instantaneous deluge of water to help neutralize any adverse reactions. Emergency showers and eyewash stations are located where people are exposed to chemicals, acids, nuclear products and fire. The water supply requirements for an emergency eyewash station are dependent on the manufacturer's requirements for the specific equipment installed.

The code does not require the installation of emergency eye wash stations and showers. This section has been provided to address them if installed. Usually, governmental agencies regulating worker and building occupant safety establish where and when these fixtures are required to be installed.

Quiz

Study Session 4
IPC Sections 401 – 411

1. Which of the following plumbing fixtures is prohibited?

    a. bidet

    b. sill cock

    c. trough urinal

    d. trench drain

    Reference _____

2. Which of the following fixtures is not required in an adult day-care facility?

    a. drinking fountain

    b. service sink

    c. lavatory

    d. bathtub

    Reference _____

3. Which of the following fixtures is required in a high school?

    a. service sink

    b. kitchen sink

    c. shower

    d. urinal

    Reference _____

4. How is the typical ratio of male to female occupants of a building determined?

    a. by the facility owner's designation, based on historical record

    b. by the design professional utilizing demographic statistical data

    c. by calculating the ratio using the total occupant load, multiplied by .5

    d. by calculating the ratio using the total occupant load, multiplied by .67

    Reference _____

5. A building classified as a business occupancy, with a occupant load of 320, is required to have a minimum of how many lavatories in the women's restroom?

    a. 2   b. 3

    c. 4   d. 5

    Reference _____

6. Which of the following is prohibited?

    a. an employee restroom where the accessible route passes through a storage room

    b. an employee restroom in a two-story mercantile occupancy located in the stockroom on the second floor

    c. a public restroom in a covered mall located 300 feet from any store in the mall

    d. a public restroom in a restaurant where the path of travel is through a kitchen corridor

    Reference _____

7. Which of the following is a *true* statement for fixtures provided within a unisex toilet room?

    a. They are permitted to be used in the computation of the minimum required fixtures.

    b. They are *not* included in the computation of the minimum required fixtures.

    c. They are required to have urinal partitions for privacy.

    d. They are *not* required to be accessible to the disabled.

    Reference _____

8. Which of the following is *not* required for public toilet facilities?

    a. signage designating sex

    b. separate facilities for each sex in a factory occupancy

    c. separate facilities for each sex where the total occupant load is 15 or less

    d. location within 500 feet of the occupied space in a mercantile occupancy

Reference _____

9. The minimum clearance for a water closet setting is _____.

    a. 30 inches from its center to a partition

    b. 15 inches from its center to a partition

    c. 21 inches center-to-center between adjacent fixtures

    d. 28 inches center-to-center between adjacent fixtures

Reference _____

10. The fixture drain from an automatic clothes washer standpipe is required to _____.

    a. be a minimum of $1^1/_2$ inches

    b. connect to a 2-inch branch drain

    c. connect to a 3-inch branch drain

    d. discharge to an approved trench drain

Reference _____

11. Which of the following is *not* permitted for the discharge of an automatic clothes washer?

    a. trench drain   b. laundry sink

    c. 2-inch standpipe   d. 2-inch branch drain

Reference _____

12. Which of the following is *not* required to have temperature limiting controls?

    a. service sink   b. whirlpool

    c. bathtub   d. bidet

Reference _____

13. Which of the following locations is prohibited for the installation of a drinking fountain?

    a. utility room
    b. restaurant

    c. restroom
    d. mortuary

Reference _____

14. A residential care facility with an occupant load of 300 is required to have at least three drinking fountains. How many bottle water dispensers can be substituted for the drinking fountains required?

    a. none
    b. one

    c. two
    d. three

Reference _____

15. Where pay toilet facilities are installed, such facilities shall be _____ the required minimum facilities.

    a. equal to
    b. in excess of

    c. computed as part of
    d. not more than half of

Reference _____

16. Joints between the drain and floor outlet fixtures or wall hung water closets shall be sealed with an approved elastomeric gasket, flange to fixture connection complying with _____ or an approved setting compound.

    a. ASME A112.4.3
    b. ASSE 1070

    c. ANSI B-125
    d. ASME A112.19.15

Reference _____

17. Joints formed where fixtures come in contact with walls and floors shall _____.

    a. be sealed
    b. be joined with setting compound

    c. meet ASME A112.19.9 M
    d. prevent ponding

Reference _____

18. The waste outlet of a bathtub shall be _____.

    a. a minimum of 2 inches in diameter

    b. equipped with a removable strainer

    c. equipped with an overflow

    d. equipped with an approved stopper

Reference _____

19. Water supply lines and fittings for every plumbing fixture shall be installed so as to prevent _____.

    a. condensation   b. backflow

    c. vibration   d. expansion

Reference _____

20. Separate facilities for each sex shall not be required in mercantile occupancies with a total occupant load of _____ or less.

    a. 50   b. 60

    c. 70   d. 80

Reference _____

21. The path of travel to required toilet facilities for an employee in a factory shall not exceed _____ feet, unless approved by the code official.

    a. 200   b. 250

    c. 300   d. 500

Reference _____

22. The path of travel to required toilet facilities for an employee in a covered mall shall not exceed _____ feet.

    a. 200   b. 250

    c. 300   d. 500

Reference _____

23. There shall be at least _____ inches of clearance in front of the water closet to any wall, fixture or door.

      a. 12   b. 18

      c. 21   d. 24

    Reference _____

24. Bidets shall be equipped with a device conforming to ASSE 1070 to limit discharge water to a maximum temperature of _____.

      a. 105°F   b. 110°F

      c. 120°F   d. 130°F

    Reference _____

25. Waste connections are not required for _____.

      a. laundry trays

      b. emergency eyewash stations

      c. automatic clothes washers

      d. floor drains

    Reference _____

2006 IPC Sections 412 – 427
Fixtures, Faucets and Fixture Fittings II

OBJECTIVE: To develop an understanding of requirements for dimensions, access and other provisions related to floor drains, food waste grinders, lavatories, showers, urinals, water closets and health care fixtures.

REFERENCE: Sections 412 through 427, 2006 *International Plumbing Code*

KEY POINTS:
- Which standards apply to a floor drain?
- What is the minimum size for a floor drain?
- Where are floor drains required?
- Which standards apply to food waste grinders?
- What is required to prevent the discharge of large particles from a food waste grinder into the drainage system?
- Which standards apply to lavatories?
- What is the minimum waste outlet size for a lavatory?
- What standards apply to prefabricated showers and shower compartments?
- What code section regulates shower valves?
- How are the supply risers to be attached to the structure?
- Unless prefabricated receptors are used, what is required for the floors under the shower?
- What is the minimum thickness for PVC flashing sheets?
- Which standards apply to sinks?
- Which standards apply to water closets?
- Which standards apply to whirlpool bathtubs?
- What locations are prohibited as a location for an ice machine in a hospital?

Code Text: *Floor drains shall have removable strainers. The floor drain shall be constructed so that the drain is capable of being cleaned. Access shall be provided to the drain inlet.*

Discussion and Commentary: The actual floor drain body may or may not have an integral trap. All floor drain installations are required to be trapped. The strainer may be flush with the floor surface or may have a dome grate. The free area of the strainer or grate must not be less than the transverse area of connecting pipe. Most manufacturers provide this information in the product literature. All floor drains must be installed or constructed such that they can be properly cleaned, ensuring that the drain will continue to function properly.

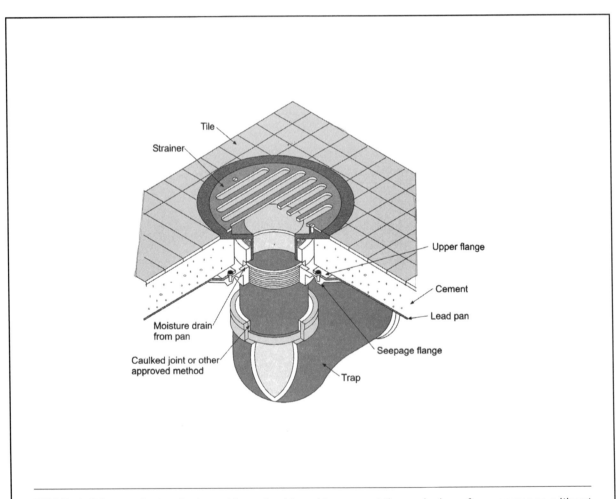

TRAP: A fitting or device that provides a liquid seal to prevent the emission of sewer gases without materially affecting the flow of sewage or wastewater through the trap.

Code Text: *Floor drains shall have a minimum 2-inch-diameter (51 mm) drain outlet.*

Discussion and Commentary: This section emphasizes the requirements of Table 709.1, which provides the minimum size of the drain outlet for the particular dfu value of the floor drain fixture. It should be noted that Section 412.4 requires a minimum floor drain size of 3 inches (76 mm) in public coin-operated laundries and in central washing facilities of multiple-family dwellings.

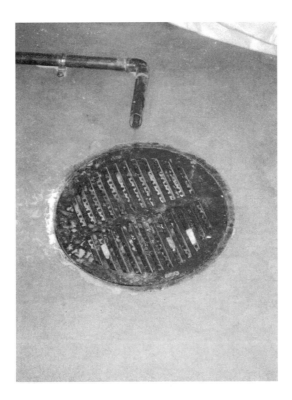

DRAINAGE FIXTURE UNIT (dfu): A measure of the probable discharge into the drainage system by various types of plumbing fixtures. The drainage fixture-unit value for a particular fixture depends on its volume rate of drainage discharge, on the time duration of a single drainage operation and on the average time between successive operations.

Code Text: *In public coin-operated laundries and in the central washing facilities of multiple-family dwellings, the rooms containing automatic clothes washers shall be provided with floor drains located to readily drain the entire floor area. Such drains shall have a minimum outlet of not less than 3 inches (76 mm) in diameter.*

Discussion and Commentary: The floor drain is required in public coin-operated laundries and in central washing facilities of multiple-family dwellings that contain laundry machines. They protect against damage from accidental spills, fixture overflows and leakage. This plumbing fixture is considered an emergency floor drain, which has a drainage fixture unit value of zero. Such fixtures do not add to the load used to compute drainage pipe sizing because their sole purpose is to serve only in the event of an emergency. It should be noted that under normal circumstances, an emergency floor drain does not receive any waste and, therefore, is required to be provided with a means to protect the trap seal from loss by evaporation.

TRAP SEAL: The vertical distance between the weir and the top of the dip of the trap.

Topic: Water Supply Required

Reference: IPC 413.4

Category: Fixtures, Faucets and Fixture Fittings

Subject: Food Waste Grinder Units

Code Text: *All food waste grinders shall be provided with a supply of cold water.*

Discussion and Commentary: A supply of cold water is necessary to act as the transport vehicle during the grinding operation. The water flushes the grinding chamber and carries the waste into the drain. If a waste grinder is used without running water, a drain blockage will result. Note that the water supply serving a commercial food waste grinder must be protected from backflow by an air gap or backflow preventer in accordance with Section 608.

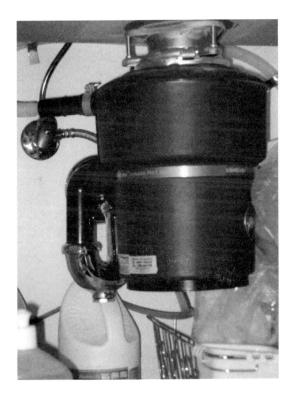

The minimum required drain size for a domestic grinder is 1.5 inches in diameter. For a commercial grinder, the minimum drain size is 2 inches in diameter (Sections 413.2 and 413.3).

Code Text: *Tempered water shall be delivered from public hand-washing facilities through an approved water temperature limiting device that conforms to ASSE 1070.*

Discussion and Commentary: The ASSE 1070 Standard is for a water temperature limiting device that is intended for this type of application. The device is a thermostatic mixing valve with a maximum temperature limit.

TEMPERED WATER: Water having a temperature range between 85°F (29°C) and 110°F (43°C).

Topic: Shower Waste Outlet

Category: Fixtures, Faucets and Fixture Fittings

Reference: IPC 417.3

Subject: Showers

Code Text: *Waste outlets serving showers shall be at least $1^{1}/_{2}$ inches (38 mm) in diameter and, for other than waste outlets in bathtubs, shall have removable strainers not less than 3 inches (76 mm) in diameter with strainer openings not less than 0.25 inch (6.4 mm) in minimum dimension. Where each shower space is not provided with an individual waste outlet, the waste outlet shall be located and the floor pitched so that waste from one shower does not flow over the floor area serving another shower. Waste outlets shall be fastened to the waste pipe in an approved manner.*

Discussion and Commentary: In gang or multiple showers, the shower room floor must slope toward the shower drains in such a manner as to prevent waste water from flowing from one shower area through another shower area. It would be undesirable and unhygienic for a bather to stand in another person's waste water.

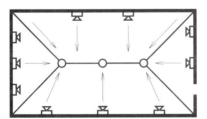

SLOPE OF SHOWER ROOM FLOOR
TO PREVENT WATER FROM DRAINING
TO ANOTHER SHOWER LOCATION

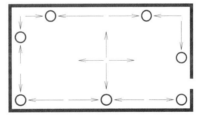

SHOWER WITH PERIMETER DRAINS

The shower drain requirements are written in performance language such that shower drains can be of different designs as long as they accomplish the code's intent.

Code Text: *All shower compartments shall have a minimum of 900 square inches (0.58 m²) of interior cross-sectional area. Shower compartments shall not be less than 30 inches (762 mm) in minimum dimension measured from the finished interior dimension of the compartment, exclusive of fixture valves, showerheads, soap dishes, and safety grab bars or rails. Except as required in Section 404, the minimum required area and dimension shall be measured from the finished interior dimension at a height equal to the top of the threshold and at a point tangent to its centerline and shall be continued to a height not less than 70 inches (1778 mm) above the shower drain outlet.*

Exception: Shower compartments having not less than 25 inches (635 mm) in minimum dimension measured from the finished interior dimension of the compartment, provided that the shower compartment has a minimum of 1,300 square inches (.838 m²) of cross-sectional area

Discussion and Commentary: A minimum of 900 square inches (0.58 mm²) of cross-sectional area is required for an average-sized adult to clean the lower body extremities by bending over. A smaller-sized shower would not provide adequate space for the user to bend over while showering. The 30-inch (726-mm) minimum dimension is based on this movement of the body.

The exception allowing a larger area of 1,300 square inches with a minimum width of 25 inches was initially intended for existing facilities replacing tubs with showers; however, the code allows this for new buildings as well.

Code Text: *The shower compartment access and egress opening shall have a minimum clear and unobstructed finished width of 22 inches (559 mm). Shower compartments required to be designed in conformance to accessibility provisions shall comply with Section 404.1.*

Discussion and Commentary: Shower compartments with sliding doors could potentially have openings as small as 17 inches. However, the minimum opening size of 22 inches would either require a larger shower compartment or elimination of the sliding doors.

Shower walls must be constructed to a minimum height of 6 feet above floor level, and be of smooth, noncorrosive and nonabsorbent waterproof materials for sanitary purposes and cleaning (Section 417.4.1).

Code Text: *Urinals shall conform to ANSI Z124.9, ASME A112.19.2M, CSA B45.1 or CSA B45.5. Urinals shall conform to the water consumption requirements of Section 604.4. Water supplied urinals shall conform to the hydraulic performance requirements of ASME A112.19.6, CSA B45.1 or CSA B45.5.*

Discussion and Commentary: There are four types of urinals: stall, blowout, siphon-jet and washdown. A stall urinal is floor mounted, with other urinals being predominantly wall hung. Both blowout and siphon-jet urinals flush completely by siphonic action. The contents of the bowl are completely evacuated during the flushing cycle, and the trap is refilled.

A washout urinal does not typically have an integral trap. Some manufacturers, however, design washout urinals with an integral trap. Neither stall nor washout urinals flush with siphon action. The flush cycle is accomplished by a combination of water exchange and dilution.

Topic: Substitution for Water Closets   **Category:** Fixtures, Faucets and Fixture Fittings
Reference: IPC 419.2   **Subject:** Urinals

Code Text: *In each bathroom or toilet room, urinals shall not be substituted for more than 67 percent of the required water closets in assembly and educational occupancies. Urinals shall not be substituted for more than 50 percent of the required water closets in all other occupancies.*

Discussion and Because a urinal is similar to a water closet in its function, it may be substituted for not more
Commentary: than 67 percent of the required water closets in assembly and educational facilities and 50 percent in other occupancies.

Assembly and Educational
67 Percent Substitution

All Other Occupancies
50 Percent Substitution

The IBC describes an assembly occupancy group as composed of buildings used for gathering, as in groups assembled for civic, social or religious purposes, or for awaiting transportation, or for recreation, food or drink consumption. Schools and day-care facilities are common examples of educational occupancies.

Topic: Public or Employee Toilet Facilities	**Category:** Fixtures, Faucets and Fixture Fittings
Reference: IPC 420.2	**Subject:** Water Closets

Code Text: *Water closet bowls for public or employee toilet facilities shall be of the elongated type.*

Discussion and Commentary: An elongated water closet bowl is required for public or employee use because it gives the user an extended area in which to eliminate human waste. The concept is to reduce the possibility of the user missing the bowl and soiling the water closet seat and surroundings. The elongated water closet bowl also has a larger water surface area, which helps keep the inside surface of the bowl clean.

An elongated water closet bowl is 2 inches (51 mm) longer than a regular bowl. The bowl extension is toward the front of the fixture. Elongated bowls are not required for various nonpublic or nonemployee uses, including hotel guestrooms and private facilities.

Topic: Access to Pump	**Category:** Fixtures, Faucets and Fixture Fittings
Reference: IPC 421.5	**Subject:** Whirlpool Bathtubs

Code Text: *Access shall be provided to circulation pumps in accordance with the fixture or pump manufacturer's installation instructions. Where the manufacturer's instructions do not specify the location and minimum size of field-fabricated access openings, a 12-inch by 12-inch (305 mm by 305 mm) minimum sized opening shall be installed to provide access to the circulation pump. Where pumps are located more than 2 feet (609 mm) from the access opening, an 18-inch by 18-inch (457 mm by 457 mm) minimum sized opening shall be installed. A door or panel shall be permitted to close the opening. In all cases, the access opening shall be unobstructed and of the size necessary to permit the removal and replacement of the circulation pump.*

Discussion and Commentary: Similar to most any other plumbing fixture, a circulation pump is also required to be accessible for maintenance and replacement purposes. The dimension of the access must be either as specified by the manufacturer or as required by the code.

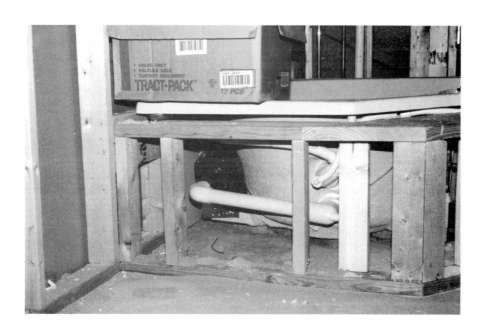

WHIRLPOOL BATHTUB: A plumbing appliance consisting of a bathtub fixture that is equipped and fitted with a circulating piping system designed to accept, circulate and discharge bathtub water upon each use.

Topic: Clinical Sink

Reference: IPC 422.6

Category: Fixtures, Faucets and Fixture Fittings

Subject: Health Care Fixtures and Equipment

Code Text: *A clinical sink shall have an integral trap in which the upper portion of a visible trap seal provides a water surface. The fixture shall be designed so as to permit complete removal of the contents by siphonic or blowout action and to reseal the trap. A flushing rim shall provide water to cleanse the interior surface. The fixture shall have the flushing and cleansing characteristics of a water closet.*

Discussion and Commentary: A clinical sink is the main special fixture used in health care facilities. It is designed to both wash a bedpan and remove its contents; therefore, the other name for a clinical sink is *bedpan washer*.

A clinical sink is like a water closet in that it has a flushing rim, which washes the interior walls of the fixture and discharges waste by siphonic action.

A hose or washing arm is provided with a clinical sink to wash the contents of the bedpan into the fixture. In some installations, a water closet functions as both a clinical sink and a water closet. The common use of the dual-purpose fixture reduces the number of plumbing fixtures required in a health-care facility.

Code Text: *Steam condensate returns from sterilizers shall be a gravity return system.*

Discussion and Commentary: Steam condensate returns from sterilizers are required to drain by gravity to reduce the likelihood of condensate backing up into a sterilizer, which could violate the sterile conditions.

Special devices and equipment requirements of Section 422 are applicable in these occupancies: nursing homes, homes for the aged, orphanages, infirmaries, first aid stations, psychiatric facilities, clinics, professional offices of dentists and doctors, mortuaries, educational facilities, surgery, dentistry, research and testing laboratories, establishments manufacturing pharmaceutical drugs and medicines, and other structures with similar apparatus and equipment classified as plumbing.

Code Text: *Hand-held showers shall conform to ASME A112.18.1 or CSA B125.1. Hand-held showers shall provide backflow protection in accordance with ASME A112.18.1 or CSA B125 or shall be protected against backflow by a device complying with ASME A112.18.3.*

Discussion and Commentary: Similar to a hose and spray assembly, it is possible for a handheld shower to be submerged in the bathtub or shower compartment base, constituting a cross connection. The handheld shower, which is illustrated below, must have adequate protection against backflow. These types of showers are commonly installed in accessible shower enclosures. Their use, however, has increased in popularity within dwelling unit showers. The referenced standards specify the requirements for backflow protection of the handheld shower.

A combination tub filler with a handheld shower creates a potential scalding hazard if the handheld unit can be wall mounted and used as a conventional shower. The requirements of Section 424.3 would apply in such cases.

Code Text: *Individual shower and tub-shower combination valves shall be balanced pressure, thermostatic or combination balanced-pressure/thermostatic valves that conform to the requirements of ASSE 1016 or CSA B125 and shall be installed at the point of use. Shower and tub-shower combination valves required by this section shall be equipped with a means to limit the maximum setting of the valve to 120°F (49°C), which shall be field adjusted in accordance with the manufacturer's instructions. In-line thermostatic valves shall not be utilized for compliance with this section.*

Discussion and Commentary: Every shower must have a control valve that is capable of protecting an individual from being scalded. These devices are also required to protect against thermal shock. Thermal shock is a change in discharge temperature great enough to produce a potentially hazardous reaction. ASSE 1016 requires control valves to protect against rapid temperature fluctuations by automatically maintaining the discharge temperature. The three types of valves available are a balanced pressure-mixing shower valve, a thermostatic-mixing shower valve and a combination mixing valve.

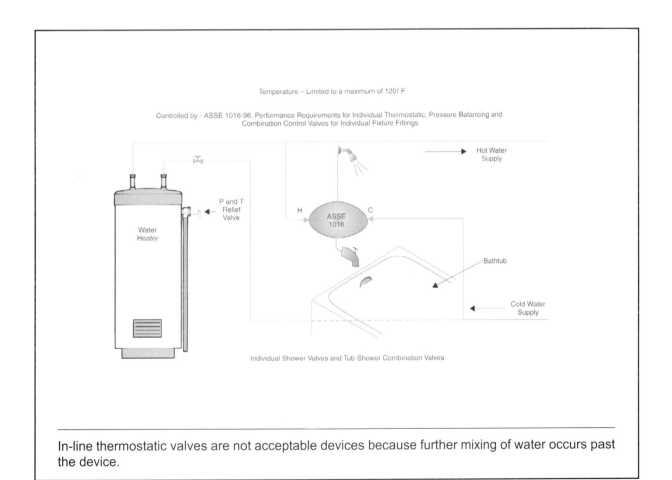

Temperature – Limited to a maximum of 120? F

Controlled by - ASSE 1016-96, Performance Requirements for Individual Thermostatic, Pressure Balancing and Combination Control Valves for Individual Fixture Fittings

Individual Shower Valves and Tub Shower Combination Valves

In-line thermostatic valves are not acceptable devices because further mixing of water occurs past the device.

Code Text: *Multiple (gang) showers supplied with a single-tempered water supply pipe shall have the water supply for such showers controlled by an approved automatic temperature control mixing valve that conforms to ASSE 1069 or CSA B125, or each shower head shall be individually controlled by a balanced-pressure, thermostatic or combination balanced-pressure/thermostatic valve that conforms to ASSE 1016 or CSA B125 and is installed at the point of use. Such valves shall be equipped with a means to limit the maximum setting of the valve to 120°F (49°C), which shall be field adjusted in accordance with the manufacturer's instructions.*

Discussion and Commentary: ASSE 1069 and CSA B125 address automatic temperature control mixing valves for gang showers.

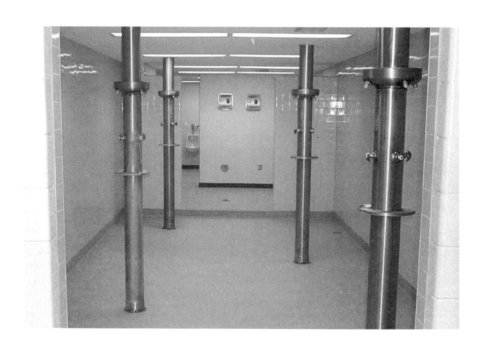

Many Canadian Standards Association (CSA) standards are referenced in the IPC as acceptable standards in lieu of ASSE or other standards.

Topic: Bathtubs and Whirlpool Valves	**Category:** Fixtures, Faucets and Fixture Fittings
Reference: IPC 424.5	**Subject:** Faucets and Other Fixture Fittings

Code Text: *The hot water supplied to bathtubs and whirlpool bathtubs shall be limited to a maximum temperature of 120°F (49°C) by a water temperature limiting device that conforms to ASSE 1070, except where such protection is otherwise provided by a combination tub/shower valve in accordance with Section 424.3.*

Discussion and Commentary: The ASSE 1070 standard is for a water temperature limiting device that is intended for this type of application. The device is a thermostatic mixing valve with a maximum temperature limit adjustment.

The ASSE 1070-04 standard provides for devices to be installed with the fixture fitting, or they can be integral to the plumbing fixture fitting supplying the water.

Code Text: *Flushometer valves and tanks shall comply with ASSE 1037. Vacuum breakers on flushometer valves shall conform to the performance requirements of ASSE 1001 or CSA B64.1.1. Access shall be provided to vacuum breakers. Flushometer valves shall be of the water-conservation type and shall not be utilized where the water pressure is lower than the minimum required for normal operation. When operated, the valve shall automatically complete the cycle of operation, opening fully and closing positively under the water supply pressure. Each flushometer valve shall be provided with a means for regulating the flow through the valve. The trap seal to the fixture shall be automatically refilled after each valve flushing cycle.*

Discussion and Commentary: There are a variety of styles of flushometer valves and tanks, all of which must be equipped with a vacuum breaker to prevent backflow. The flushometer valve operation is either a diaphragm type or a piston type. The majority of flushometer valves are diaphragm type, on account of its reliability. The working components of a flushometer valve are either exposed or concealed. When concealed, the valve must be accessible through an access opening, cover or plate.

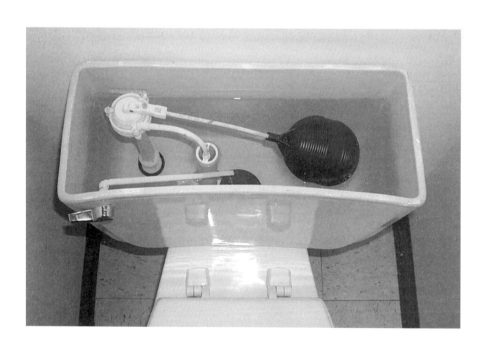

Flushometers are either manually or automatically operated. Common manual valves are activated by a hand lever, push button or foot valve. Automatic valves are activated electrically by infrared sensors or other devices that sense the presence of an individual using the fixture.

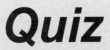

Quiz

Study Session 5
IPC Sections 412 – 427

1. Which of the following is not a requirement for water closets installed in public and employee restrooms?

    a. nonabsorbent seats
    b. elongated seats
    c. open front seats
    d. integral seats

    Reference _____

2. In each toilet room of an institutional occupancy, urinals may be substituted for no more than _____ percent of the required number of water closets?

    a. 50
    b. 33
    c. 67
    d. 75

    Reference _____

3. Which of the following is a minimum requirement for a sink?

    a. 1.25-inch waste outlet
    b. 2-inch waste outlet
    c. strainer or crossbar
    d. temperature limiting device

    Reference _____

4. Water delivered from public hand-washing lavatories shall have a temperature range between _____.

    a. 80°F and 105°F   b. 85°F and 110°F

    c. 90°F and 115°F   d. 95°F and 120°F

Reference _____

5. Unless a shower compartment is designed for accessibility, the required minimum width of the door opening is _____ inches.

    a. 22   b. 25

    c. 27   d. 30

Reference _____

6. For a shower compartment with an area of 1,300 square inches, the minimum interior dimension is _____ inches.

    a. 22   b. 25

    c. 27   d. 30

Reference _____

7. Where the manufacturer does not specify the location and minimum size of the field fabricated access opening to a whirlpool circulation pump, the minimum dimensions are _____ , provided the pump is less than 24 inches from the access opening.

    a. 8 inches x 12 inches   b. 12 inches x 12 inches

    c. 12 inches x 18 inches   d. 18 inches x 18 inches

Reference _____

8. Which of the following is approved for the control of temperature from an individual shower valve?

    a. in-line thermostatic valve

    b. ASSE 1017, temperature actuated mixing valve

    c. ASSE 1070, water-temperature limiting device

    d. ASSE 1016, combination balanced pressure/thermostatic valve

Reference _____

9. The maximum allowable water temperature supplied by an individual shower valve is

_____ .

    a. 100°F   b. 110°F

    c. 120°F   d. 140°F

Reference _____

10. Which of the following is not approved for protection of the water supply connected to a garbage can washer?

    a. reduced pressure principle backflow preventer

    b. backflow preventer with intermediate atmospheric vent

    c. double check valve assembly

    d. atmospheric vacuum breaker

Reference _____

11. In each bathroom or toilet room, urinals shall not be substituted for more than _____ percent of the required water closets in an educational occupancy.

    a. 33   b. 50

    c. 67   d. 75

Reference _____

12. The drain outlet for floor drains shall be a minimum of _____ inches in diameter.

    a. $1^1/_2$ b. 2

    c. $1^1/_4$ d. 3

Reference _____

13. Floor drains in public coin operated laundries shall have a minimum outlet cross section of not less than _____ inches in diameter.

    a. 4   b. 3

    c. 2   d. $1^1/_2$

Reference _____

14. Domestic food waste grinders shall be connected to a drain of not less than _____ inches in diameter.

    a. 2   b. $1^{1}/_{4}$

    c. 3   d. $1^{1}/_{2}$

Reference _____

15. A shower liner of 0.040-inch thick PVC sheet must turn up on all sides at least _____ inches above the threshold level.

    a. 2   b. 3

    c. 1.5   d. 2.5

Reference _____

16. Flushometer valves shall be of the _____ type.

    a. water-pressure   b. diaphragm

    c. water-conservation   d. antisiphon

Reference _____

17. The opening of the overflow pipe in a flush tank shall be located above the _____ of the water closet or above a secondary overflow in the flush tank.

    a. trap seal   b. flood level rim

    c. fill valve   d. backflow preventer

Reference _____

18. A _____ shall have the flushing and cleansing characteristics of a water closet.

    a. clinical sink   b. urinal

    c. bidet   d. floor sink

Reference _____

19. A _____ shall not serve more than one fixture.

    a. transfer valve   b. flushing device

    c. temperature limiting device d. vacuum breaker

Reference _____

20. The fill valve backflow preventer in a flush tank shall be located at least_____ inch(es) above the full opening of the overflow pipe.

      a. $\frac{1}{2}$ b. 1

      c. $1\frac{1}{2}$ d. 2

Reference _____

21. Steam condensate return piping from sterilizers shall be _____.

      a. stainless steel piping   b. plastic piping

      c. a gravity return system   d. trapped

Reference _____

22. Lavatories shall have waste outlets not less than _____ inch(es) in diameter.

      a. $1\frac{1}{2}$ b. $1\frac{1}{4}$

      c. 1   d. 2

Reference _____

23. Removable strainers at waste outlets of showers shall be not less than _____ inches in diameter.

      a. 3   b. 2

      c. 5   d. 4

Reference _____

24. Where the manufacturer does not specify the location and minimum size of the field fabricated access opening to a whirlpool circulation pump and the pump is more than 24 inches from the access opening, the minimum opening dimensions are _____.

      a. 18 inches x 24 inches   b. 12 inches x 12 inches

      c. 12 inches x 18 inches   d. 18 inches x 18 inches

Reference _____

25. A water temperature limiting device shall be provided to limit the hot water provided to a whirlpool bathtub to a maximum of _____.

    a. 110°F   b. 115°F

    c. 120°F   d. 125°F

Reference _____

2006 IPC Chapter 5
Water Heaters

OBJECTIVE: To develop an understanding of the requirements for safe installation of water heaters. To develop an understanding of provisions of the code that apply to water heaters used for space heating.

REFERENCE: Chapter 5, 2006 *International Plumbing Code*

KEY POINTS:
- What areas/elements are covered by the provisions of Chapter 5?
- What is the maximum temperature of the water when the water heater is a part of a space heating system?
- In what condition is the water to be maintained throughout the domestic hot water system?
- Where are drain valves for a water heater to be located?
- Which standards apply to drain valves for water heaters?
- What is the purpose of drain valves on water heaters?
- When intended for domestic uses, the domestic hot water temperature is limited to what maximum temperature?
- Domestic tanks and water heaters are required to have what information permanently attached to them?
- Where are the markings required to be located on a water heater?
- Are all domestic hot water supply systems required to be equipped with temperature controls?
- What type of water heater is prohibited from a room used as a plenum?
- What type of flooring is required?
- Where is the service space required, and what is the required size?
- What is the minimum access opening?
- What device is required on the cold water branch line to the water heater?
- Where is the shut-off valve for a water heater to be located?
- How is proper circulation of water through the water heater obtained?
- Pipes/tubes that draw from the water heater are required to comply with what provisions of the code?
- Energy cutoff devices are required on what type of water heaters?

KEY POINTS:
(Cont'd)

- Does the energy cutoff device on a water heater serve as a temperature and pressure relief valve?
- When are storage water heaters required to have relief valves?
- Which standard applies to water heater temperature and pressure relief valves?
- Does the code allow the temperature and pressure relief valve to serve as a means of controlling thermal expansion?
- Isolation valves serving water heaters are to be installed in what location?
- Where are temperature and pressure relief valves to be located for a water heater tank?
- Are valves permitted between the water heater and the storage tank?
- Are check valves or shutoff valves permitted between a relief valve and the heater or tank?
- When are pans required under water heaters?
- What is the minimum depth of the pan?
- What limitation is placed on heat loss for water heaters?
- What is the design ambient temperature for determining heat loss for water heaters?

Topic: Water Heater as Space Heater	**Category:** Water Heaters
Reference: IPC 501.2	**Subject:** General Provisions

Code Text: *Where a combination potable water heating and space heating system requires water for space heating at temperatures higher than 140°F (60°C), a master thermostatic mixing valve complying with ASSE 1017 shall be provided to limit the water supplied to the potable hot water distribution system to a temperature of 140°F (60°C) or less. The potability of the water shall be maintained throughout the system.*

Discussion and Commentary: When a water heater has a dual purpose of supplying hot water and serving as a heat source for a hot water space heating system, the maximum outlet water temperature for the potable hot water distribution system is limited to 140°F (60°C). A master thermostatic mixing valve conforming to ASSE 1017 must be provided to limit the water temperature to 140° (60°C) or less. These valves are used extensively in applications for domestic service to mix hot and cold water to reduce high service water temperature to the building distribution system. These devices are not intended for final temperature control at fixtures and appliances (see ASSE 1016).

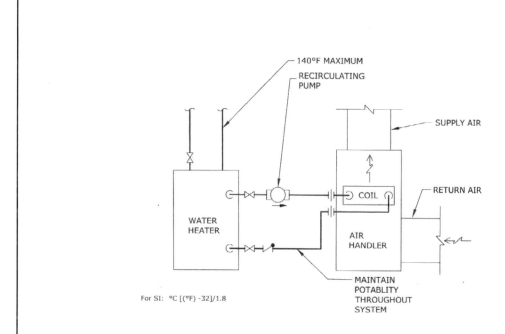

For SI: °C [(°F) -32]/1.8

WATER-TO-AIR HEAT EXCHANGER

Chemicals of any type must not be added to the heating system, as this would directly contaminate the potable water supply. Protection of the potable water supply must be in accordance with Section 608.

Code Text: *Drain valves for emptying shall be installed at the bottom of each tank-type water heater and hot water storage tank. Drain valves shall conform to ASSE 1005.*

Discussion and Commentary: A water heater must be capable of being drained to facilitate service, sediment removal, repair or replacement. Drain valves must be constructed and tested in accordance with ASSE 1005, which requires the valves to have an inlet of $^3/_4$-inch (19.1 mm) nominal iron pipe size, the outlet to be equipped with a standard $^3/_4$-inch (19.1 mm) male hose thread and a straight-through waterway of not less than $^1/_2$-inch (12.7 mm) diameter.

WATER HEATER: Any heating appliance or equipment that heats potable water and supplies such water to the potable hot water distribution system.

Code Text: *Water heaters and storage tanks shall be located and connected so as to provide access for observation, maintenance, servicing and replacement.*

Discussion and Commentary: Because water heaters and the potable water connections require routine inspections, maintenance, repairs and possible replacement, access is required. Additionally, access recommendations or requirements are usually stated in the manufacturers' installation instructions. Thus, the provisions stated herein are intended to supplement the manufacturers' installation instructions. The intent is to provide access to all components that require observation, inspection, adjustment, servicing, repair and replacement. Access is also necessary to conduct operating procedures such as start-up or shutdown.

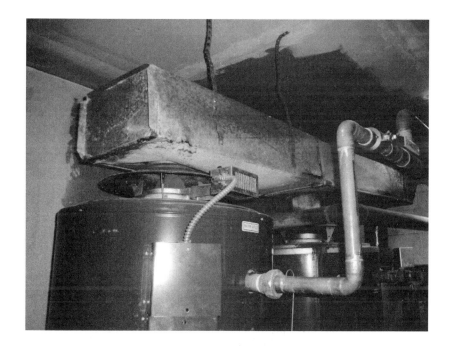

The code defines access as being able to be reached but which first may require the removal of a panel, door or similar obstruction.

Code Text: *All water heaters shall be third-party certified.*

Discussion and Commentary: Water heaters must be third-party certified by an approved agency as complying with the applicable and appropriate national standards. Some examples of standards that are used as a basis for testing and certification of water heaters include ANSI Z21.10.1 and ANSI Z21.10.3 for gas-burning water heaters, UL 174 and UL 1453 for domestic electric water heaters, UL 1453 for commercial electric water heaters and UL 732 for oil-burning water heaters.

Reliance upon the certification process to indicate the performance characteristics of the water heater is a fundamental principle of the code. The presence of a certification mark is part of the total information that the code official needs to consider in the approval of the water heater.

Code Text: *The temperature of water from tankless water heaters shall be a maximum of 140°F (60°C) when intended for domestic uses. This provision shall not supersede the requirement for protective shower valves in accordance with Section 424.3.*

Discussion and Commentary: The intent of this section is to prevent excessively high water temperatures from reaching plumbing fixtures. Tankless heaters do not have a storage capacity and are often called *instantaneous* because the water is heated at the same time and at the same rate it is used. Instantaneous heaters can discharge an uncertain range of temperatures at any given time, depending on the use. Therefore, some form of temperature control is necessary to protect the user against exposure to excessively hot water discharged from domestic fixtures such as lavatories, kitchen sinks, tubs and laundry trays. This is accomplished with a tempering valve adjusted to deliver water at a maximum temperature of 140°F (60°C), or equipping the heater with a temperature-limiting device or thermostat that has a maximum setting of 140°F (60°C).

424.3 Individual shower valves. Individual shower and tub-shower combination valves shall be balanced-pressure, thermostatic or combination balanced-pressure/thermostatic valves that conform to the requirements of ASSE 1016 or CSA B125 and shall be installed at the point of use. Shower and tub-shower combination valves required by this section shall be equipped with a means to limit the maximum setting of the valve to 120°F (49°C), which shall be field adjusted in accordance with the manufacturer's instructions. In-line thermostatic valves shall not be utilized for compliance with this section.

When a tankless water heater is used to supply hot water to a shower and tub/shower combination, the maximum outlet temperature must be controlled at 120°F (49°C) in accordance with Section 424.3.

Code Text: *Storage tanks and water heaters installed for domestic hot water shall have the maximum allowable working pressure clearly and indelibly stamped in the metal or marked on a plate welded thereto or otherwise permanently attached. Such markings shall be in an accessible position outside of the tank so as to make inspection or reinspection readily possible.*

Discussion and Commentary: Water heaters, like all pressure vessels, must be able to withstand the working pressures to which they will be subjected. The maximum working pressure of the water heater must be known so that a properly rated relief valve not exceeding the manufacturer's rated working pressure can be installed. The working pressure must be marked in such a way that it is permanent, not susceptible to damage and legible for the life of the water heater. Therefore, if the relief valve requires replacement, a properly sized valve can be reinstalled.

Allstar Testing

Tested at 300 PSI

Max. operating pressure 150 PSI

Capacity 40 gallons

The marking on the water heater must be located so that it is readily accessible during inspecting and servicing.

Code Text: *Water heaters shall be installed in accordance with the manufacturer's installation instructions. Oil-fired water heaters shall conform to the requirements of this code and the* International Mechanical Code. *Electric water heaters shall conform to the requirements of this code and provisions of the* Electrical Code. *Gas-fired water heaters shall conform to the requirements of the* International Fuel Gas Code.

Discussion and Commentary: By placing the mechanical equipment in the attic, it is possible to better utilize the building's floor area. However, it is important that such equipment be accessible for service or removal. In addition to a properly sized access opening, a continuous passageway of solid flooring must be provided where needed to reach the appliance. A minimum 30-inch by 30-inch service area is required in front of any appliance access.

CHAPTER 13

REFERENCED STANDARDS

This chapter lists the standards that are referenced in various sections of this document. The standards are listed herein by the promulgating agency of the standard, the standard identification, the effective date and title, and the section or sections of this document that reference the standard. The application of the referenced standards shall be as specified in Section 102.8.

ANSI

American National Standards Institute
25 West 43rd Street, Fourth Floor
New York, NY 10036

Standard Reference Number	Title	Referenced in code section number
Z4.3—95	Minimum Requirements for Nonsewered Waste-Disposal Systems	311.1
Z21.22—99 (R2003)	Relief Valves for Hot Water Supply Systems with Addenda Z21.22a-2000(R2003) and Z21.22b-2001(R2003)	504.2, 504.5
Z124.1—95	Plastic Bathtub Units	407.1
Z124.2—95	Plastic Shower Receptors and Shower Stalls	417.1
Z124.3—95	Plastic Lavatories	416.1, 416.2
Z124.4—96	Plastic Water Closet Bowls and Tanks	420.1
Z124.6—97	Plastic Sinks	415.1, 418.1
Z124.9-94	Plastic Urinal Fixtures	419.1

The manufacturer's instructions are thoroughly evaluated by the third-party certification agency to ensure safe installation. The certifying agency can require the manufacturer to alter, delete or add information in the installation instructions as necessary to achieve compliance with the applicable standards and code requirements.

Topic: Seismic Supports

Category: Water Heaters

Reference: IPC 502.4

Subject: Installation

Code Text: *Where earthquake loads are applicable in accordance with the* International Building Code, *water heater supports shall be designed and installed for the seismic forces in accordance with the* International Building Code.

Discussion and Commentary: Chapter 16 of the *International Building Code* (IBC) contains determination of seismic loads based on Seismic Design Category. IPC Section 308.2 requires piping supports to be designed to resist seismic loads. Similarly, this section requires water heater supports to be designed to resist the same seismic loads in accordance with the IBC.

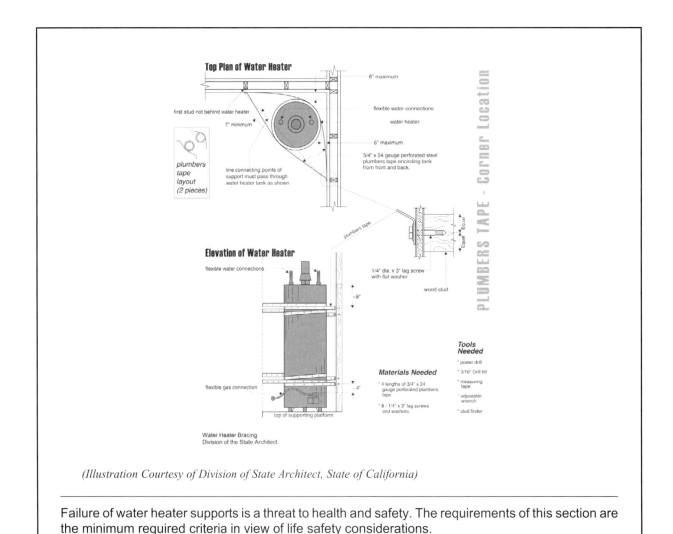

(Illustration Courtesy of Division of State Architect, State of California)

Failure of water heater supports is a threat to health and safety. The requirements of this section are the minimum required criteria in view of life safety considerations.

Code Text: *The cold water branch line from the main water supply line to each hot water storage tank or water heater shall be provided with a valve, located near the equipment and serving only the hot water storage tank or water heater. The valve shall not interfere or cause a disruption of the cold water supply to the remainder of the cold water system. The valve shall be provided with access on the same floor level as the water heater served.*

Discussion and Commentary: This section requires a valve located near the water heater to be installed in the cold water branch line from the main water supply. This section also provides requirements for separate water heater and storage tank hot water systems. Valves are needed to isolate each hot water storage tank or water heater from the water distribution system to facilitate service, repair and replacement and to allow for emergency shutoff in the event of a failure. The shutoff valve must be adjacent to the hot water storage tank or water heater, conspicuously located and within reach to permit it to be easily located and operated in the event of an emergency. The valve is only to serve the hot water storage tank or water heater, thereby not disrupting the flow of the cold water supply to other portions of the water distribution system when the hot water storage tank or water heater is taken out of service.

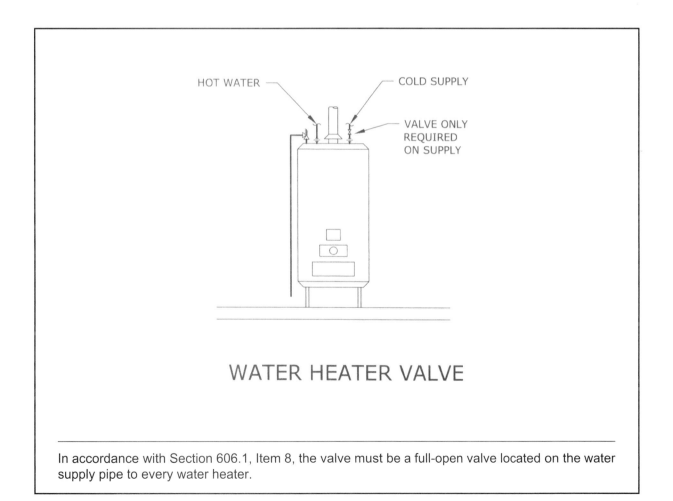

HOT WATER

COLD SUPPLY

VALVE ONLY
REQUIRED
ON SUPPLY

WATER HEATER VALVE

In accordance with Section 606.1, Item 8, the valve must be a full-open valve located on the water supply pipe to every water heater.

Code Text: *An approved means, such as a cold water "dip" tube with a hole at the top or a vacuum relief valve installed in the cold water supply line above the top of the heater or tank, shall be provided to prevent siphoning of any storage water heater or tank.*

Discussion and Commentary: This section establishes the requirements for the approval and installation of safety devices on water heaters. Water heaters are designed to operate only when they are full of water. If some or all of the water is siphoned out of the tank during an interruption in the cold water supply, damage to the water heater from overheating could occur. Typically, water heater designs include a cold water *dip* tube, which directs the incoming cold water to the bottom of the tank. At the top of the tank, a hole is provided in the dip tube to prevent water from being siphoned from the tank through the tube. ANSI Z21.10.1 and UL 174 both require the dip tube to be provided with a hole located within 6 inches (152 mm) of the top of the tank.

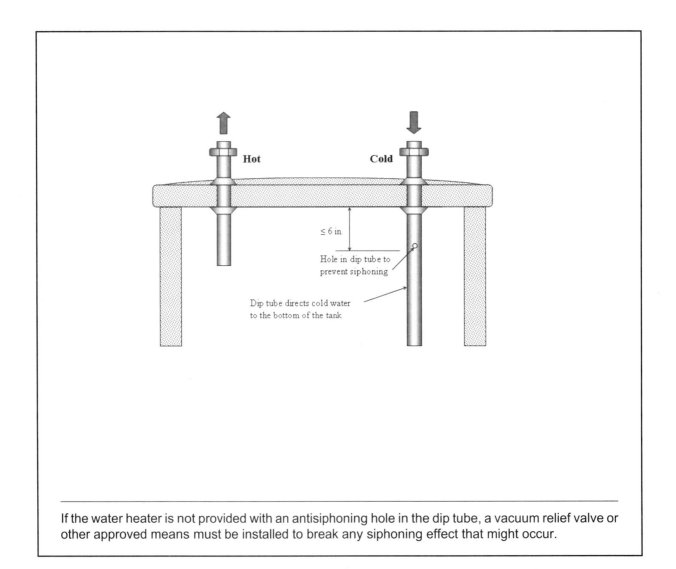

If the water heater is not provided with an antisiphoning hole in the dip tube, a vacuum relief valve or other approved means must be installed to break any siphoning effect that might occur.

Code Text: *A means for disconnecting an electric hot water supply system from its energy supply shall be provided in accordance with the* Electrical Code. *A separate valve shall be provided to shut off the energy fuel supply to all other types of hot water supply systems.*

Discussion and Commentary: This section parallels the *Electrical Code*, the *International Mechanical Code* (IMC) and the *International Fuel Gas Code* (IFGC) by requiring all water heaters to be capable of having the fuel or power supply turned off without affecting other appliances, equipment or systems. The shutoff valves for fuel-fired water heaters and disconnects for electric water heaters are necessary to allow for service, repairs and temporary and emergency shutdown.

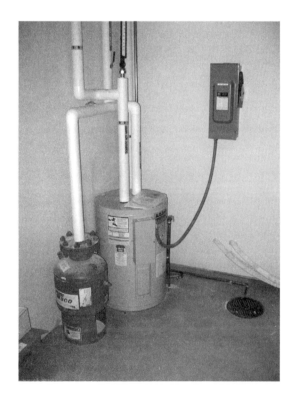

Emergency shut-down of water heaters is critically important, as the potential of explosion and severe injury are present in any enclosed system that is heated.

Code Text: *All storage water heaters operating above atmospheric pressure shall be provided with an approved, self-closing (levered) pressure relief valve and temperature relief valve or combination thereof. The relief valve shall conform to ANSI Z21.22. The relief valve shall not be used as a means of controlling thermal expansion.*

Discussion and Commentary: A combination temperature and pressure relief valve, or separate temperature relief and pressure relief valves, protect the water heater from possible explosion. Every water heater except instantaneous point-of-use heaters must have both temperature and pressure relief protection.

Water heaters installed without temperature and pressure relief valve protection can produce devastating explosions and have been responsible for deaths and property damage. Relief valves also play an important role in protecting aging tanks, insofar as corrosive elements in water cause a high rate of tank deterioration and weakening.

Pressure relief valves are designed to relieve excessive pressures that can develop in a closed vessel, tank or system. Temperature relief valves are designed to open in response to excessive temperatures and discharge heated water to limit the temperature of the water in the vessel, tank or system.

Code Text: *Such valves shall be installed in the shell of the water heater tank. Temperature relief valves shall be so located in the tank as to be actuated by the water in the top 6 inches (152 mm) of the tank served. For installations with separate storage tanks, the valves shall be installed on the tank and there shall not be any type of valve installed between the water heater and the storage tank. There shall not be a check valve or shut off valve between a relief valve and the heater or tank served.*

Discussion and Commentary: Water heaters are typically provided with factory-installed openings that are properly located to receive a relief valve device. Such openings must be used for that purpose. ANSI Z21.10.1 and UL 174 both require a temperature and pressure relief valve to be installed in a location specified by the manufacturer or for the water heater to be provided with a temperature and pressure relief valve.

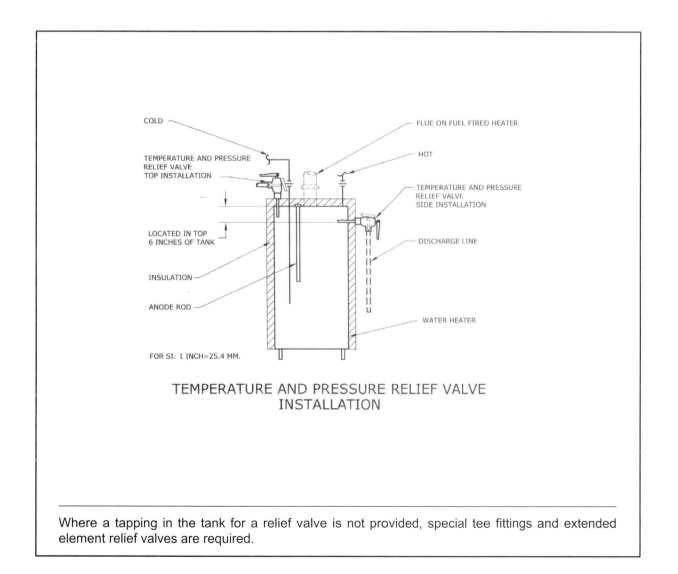

COLD

FLUE ON FUEL FIRED HEATER

TEMPERATURE AND PRESSURE
RELIEF VALVE
TOP INSTALLATION

HOT

TEMPERATURE AND PRESSURE
RELIEF VALVE
SIDE INSTALLATION

LOCATED IN TOP
6 INCHES OF TANK

DISCHARGE LINE

INSULATION

ANODE ROD

WATER HEATER

FOR SI: 1 INCH=25.4 MM.

TEMPERATURE AND PRESSURE RELIEF VALVE
INSTALLATION

Where a tapping in the tank for a relief valve is not provided, special tee fittings and extended element relief valves are required.

Code Text: *The discharge piping serving a pressure relief valve, temperature relief valve or combination thereof shall:*

1. *Not be directly connected to the drainage system.*
2. *Discharge through an air gap located in the same room as the water heater.*
3. *Not be smaller than the diameter of the outlet of the valve served and shall discharge full size to the air gap.*
4. *Serve a single relief device and shall not connect to piping serving any other relief device or equipment.*
5. *Discharge to the floor, to an indirect waste receptor or to the outdoors. Where discharging to the outdoors in areas subject to freezing, discharge piping shall be first piped to an indirect waste receptor through an air gap located in a conditioned area.*
6. *Discharge in a manner that does not cause personal injury or structural damage.*
7. *Discharge to a termination point that is readily observable by the building occupants.*
8. *Not be trapped.*
9. *Be installed so as to flow by gravity.*
10. *Not terminate more than 6 inches (152 mm) above the floor or waste receptor.*
11. *Not have a threaded connection at the end of such piping.*
12. *Not have valves or tee fittings.*
13. *Be constructed of those materials listed in Section 605.4 or materials tested, rated and approved for such use in accordance with ASME A112.4.1.*

Discussion and Commentary: The requirements for relief valve discharge piping have been itemized for clear understanding of the provisions. Each item describes the limitations or the needed safety features. In item 4, a discharge pipe must serve only one relief device so that the source of the problem can be immediately identified.

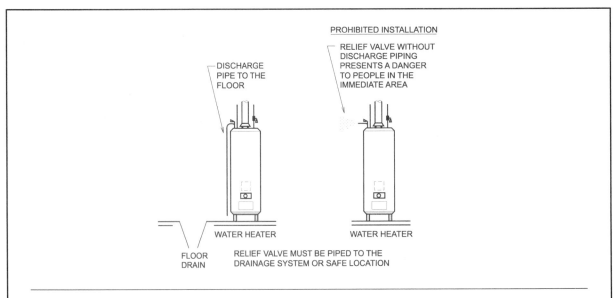

When a discharge takes place, the code intends to protect against personal injury and structural damage, but not necessarily other property damage.

Topic: Pan Size and Drain

Reference: IPC 504.7.1

Category: Water Heaters

Subject: Safety Devices

Code Text: *The pan shall be not less than 1.5 inches (38 mm) deep and shall be of sufficient size and shape to receive all dripping or condensate from the tank or water heater. The pan shall be drained by an indirect waste pipe having a minimum diameter of 0.75 inch (19 mm). Piping for safety pan drains shall be of those materials listed in Table 605.4.*

Discussion and Commentary: The purpose of the pan is to prevent water from damaging the area surrounding the water heater or tank. The pan must be of sufficient size and shape so as to catch all dripping water or condensate. The pan must be drained by an indirect waste pipe having a minimum diameter of $^3/_4$ inch (19.1 mm).

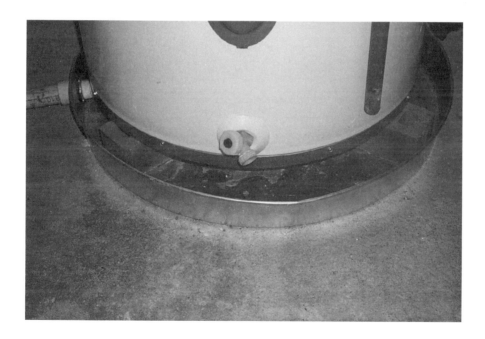

INDIRECT WASTE PIPE: A waste pipe that does not connect directly with the drainage system, but that discharges into the drainage system through an air break or air gap into a trap, fixture, receptor or interceptor.

Topic: Pan Drain Termination	**Category:** Water Heaters
Reference: IPC 504.7.2	**Subject:** Safety Devices

Code Text: *The pan drain shall extend full-size and terminate over a suitably located indirect waste receptor or floor drain or extend to the exterior of the building and terminate not less than 6 inches (152 mm) and not more than 24 inches (610 mm) above the adjacent ground surface.*

Discussion and Commentary: The pan drain must not be reduced in size over its length, as any reduction will act as a restriction and impede the discharge. The pan drain must terminate to an indirect waste receptor, floor drain or extend to the exterior of the building. An air gap or air break must be provided to prevent backflow when the pan drain terminates into an indirect waste receptor or a floor drain. When the pan drain terminates to the exterior of the building, the termination must not be less than 6 inches (152 mm) nor more than 24 inches (610 mm) above the adjacent ground surface.

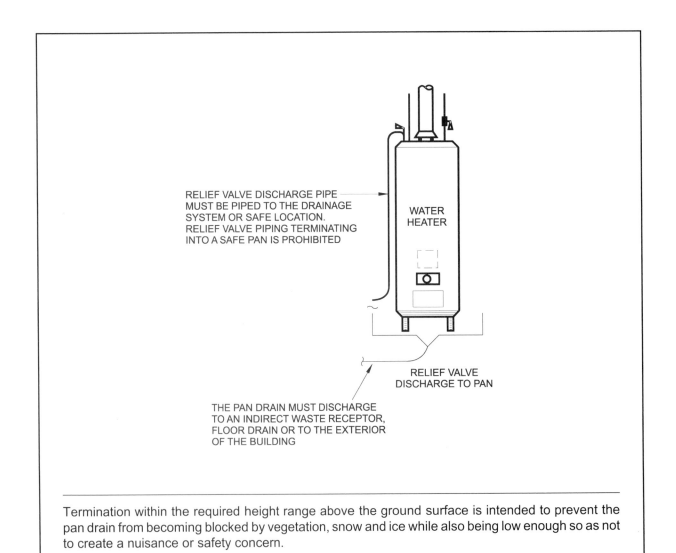

RELIEF VALVE DISCHARGE PIPE MUST BE PIPED TO THE DRAINAGE SYSTEM OR SAFE LOCATION. RELIEF VALVE PIPING TERMINATING INTO A SAFE PAN IS PROHIBITED

WATER HEATER

RELIEF VALVE DISCHARGE TO PAN

THE PAN DRAIN MUST DISCHARGE TO AN INDIRECT WASTE RECEPTOR, FLOOR DRAIN OR TO THE EXTERIOR OF THE BUILDING

Termination within the required height range above the ground surface is intended to prevent the pan drain from becoming blocked by vegetation, snow and ice while also being low enough so as not to create a nuisance or safety concern.

Quiz

Study Session 6
IPC Chapter 5

1. All of the following are minimum requirements of a water heater installation, *except* _____ .

    a. a valve on the cold water supply to the water heater

    b. a drain valve at the bottom of each tank-type water heater

    c. a means to prevent siphoning of the tank in the event of a negative supply pressure

    d. a means to provide continuous circulation from the piping system to the heater

    Reference _____

2. Which of the following is required on a bottom fed water heater?

    a. dip tube   b. circulating pump

    c. vacuum relief valve   d. thermal expansion device

    Reference _____

3. What is the maximum pressure setting for a temperature and pressure relief valve?

    a. 125 psi

    b. 150 psi

    c. rated working pressure of the tank

    d. lesser of the tank working pressure or 150 psi

    Reference _____

4. A combination temperature and pressure relief valve is required to be installed so that it is actuated by_____ .

    a. a thermostatic cut-off

    b. a loss of water supply pressure

    c. the temperature in the top 6 inches of the tank

    d. the loss of electrical energy or the fuel supply

Reference _____

5. The maximum temperature setting of a temperature and pressure relief valve is _____ .

    a. 120°F   b. 140°F

    c. 180°F   d. 210°F

Reference _____

6. When is a water heater required to be installed in a pan?

    a. when required by the code official

    b. when installed in an attic

    c. when tank leakage would cause damage to the structure

    d. when required by the manufacturer

Reference _____

7. The drain pipe of a water heater pan shall be not less than _____ inch in diameter.

    a. 0.75   b. 0.5

    c. 0.325   d. 1.0

Reference _____

8. The temperature and pressure relief valve drain shall not _____ .

    a. discharge to the floor

    b. discharge to the outdoors

    c. discharge to an indirect waste receptor

    d. connect to the drainage system

Reference _____

9. All of the following are prohibited in a temperature and pressure relief valve drain, except _____.

    a. traps   b. valves

    c. elbow fittings   d. threaded ends

Reference _____

10. The relieving capacity of a temperature relief valve shall not be less than the _____ of the water heater.

    a. heat input   b. heat output

    c. temperature limit   d. pressure limit

Reference _____

11. The termination of a temperature and pressure relief valve discharge pipe shall be _____.

    a. not less than 6 inches above the floor

    b. through an air break

    c. to a floor drain

    d. readily observable

Reference _____

12. What is the maximum heat loss allowed for an unfired hot water storage tank?

    a. an amount not to exceed the hourly recovery rate of the water heater supply

    b. an amount not to exceed one-half the hourly recovery rate of the water heater

    c. 15 Btu/h per square foot of external tank surface

    d. 15 Btu/h per cubic foot of storage capacity

Reference _____

13. What is the minimum diameter of a safety pan drain?

    a. $^1/_2$ inch   b. $^3/_8$ inch

    c. $^3/_4$ inch   d. 1 inch

Reference _____

14. The maximum allowable water temperature from a tankless-type water heater intended for domestic use is _____ .

    a. 120°F   b. 140°F

    c. 160°F   d. 180°F

Reference _____

15. The valve on the cold water supply to a water heater shall not _____ .

    a. be a full-open type

    b. serve any other fixture

    c. be adjacent to the water heater

    d. be located more than 24 inches from the water heater

Reference _____

16. A water heater requires _____ certification.

    a. code official   b. third-party

    c. manufacturer   d. design professional

Reference _____

17. Which of the following is required to limit the water temperature supplied by a combination water heater and space heating appliance?

    a. temperature limiting device conforming to ASSE 1070

    b. individual thermostatic valve conforming to ASSE 1016

    c. master thermostatic mixing valve conforming to ASSE 1017

    d. temperature-actuated, flow reduction device conforming to ASSE 1062

Reference _____

18. The _____ shall be clearly marked on storage tank water heaters installed for domestic hot water.

    a. pressure test agency   b. maximum test pressure

    c. maximum working pressure d. maximum relief pressure

Reference _____

19. What is the minimum access opening size required for a water heater installed in an attic?

    a. 20 inches by 30 inches   b. 22 inches by 30 inches.

    c. 24 inches by 30 inches   d. 30 inches by 30 inches

Reference _____

20. An electric water heater requires a _____.

    a. drain pan   b. floor drain

    c. vacuum relief valve   d. disconnecting means

Reference _____

21. A pan for a water heater shall not be less than _____ inches deep.

    a. $1^1/_2$ b. 2

    c. $3^1/_2$ d. 4

Reference _____

22. The pan drain shall terminate not more than _____ inches above the adjacent ground surface.

    a. 6   b. 12

    c. 18   d. 24

Reference _____

23. Temperature relief valves shall be so located in the tank as to be actuated by the water in the _____ inches.

    a. bottom 6   b. bottom 12

    c. top 6   d. top 12

Reference _____

24. A master thermostatic mixing valve complying with ASSE 1017 shall be provided to limit the water supplied to the potable hot water distribution system to a temperature of _____ or less.

    a. 120°F   b. 140°F

    c. 160°F   d. 180°F

Reference _____

25. A passageway to a water heater installed in an attic shall have continuous solid flooring not less than _____ inches wide.

    a. 22   b. 24

    c. 30   d. 36

Reference _____

2006 IPC Sections 601 – 605
Water Supply and Distribution I

OBJECTIVE: To develop an understanding of the code provisions related to water supply sources, water supply pipe locations, pumps, joints and fittings.

REFERENCE: Sections 601 through 605, 2006 *International Plumbing Code*

KEY POINTS:
- Are individual water supplies exempt from code requirements?
- What types of structures are required to have a supply of potable water?
- Is potable water required to all plumbing fixtures?
- What determines the size of the pipe for the water supply?
- What must be provided if the water pipe crosses a sewer line and is closer than 5 feet?
- What governs the design of water systems?
- What table provides the design criteria at the fixture supply pipe outlets?
- Are there any exceptions to the general provisions for maximum consumption for water closets?
- Where pressure may fluctuate, what is used as the design pressure?
- At what pressure does the code require a pressure-reducing valve?
- How is water hammer to be prevented?
- Which table applies to the manifold sizing, and what does the total gallons per minute represent?
- What identification is needed for individual shutoff valves?
- Is access required to manifolds?
- What is the minimum working pressure for water pipe or tubing installed underground and outside of the structure?
- Which standard applies to a dual check-valve-backflow preventer?
- What is the minimum pressure rating for hot water distribution pipe and tubing?
- What limitations are placed on fittings to prevent retarding or obstructing the water flow in the pipe?
- Flexible water connectors exposed to continuous pressure are required to comply with what standard?

KEY POINTS:
(Cont'd)

- Valves are required to comply with what two criteria?
- What type of seal is required for mechanical joints on water pipe?
- Where are solvent-cement joints allowed?
- What type of coupling and seal is required for asbestos-cement pipe?
- What regulates the installation of mechanical joints?

Topic: Solar Energy Utilization	**Category:** Water Supply and Distribution
Reference: IPC 601.2	**Subject:** General Provisions

Code Text: *Solar energy systems used for heating potable water or using an independent medium for heating potable water shall comply with the applicable requirements of this code. The use of solar energy shall not compromise the requirements for cross connection or protection of the potable water supply system required by* the International Plumbing Code.

Discussion and Commentary: Solar heating systems consist of two basic types: direct connection and indirect connection. In a direct connection system, the heat transfer fluid is potable water. The *International Mechanical Code* (IMC) contains information on the design and installation of solar heating systems. The potable water source must be protected from possible contamination from piping and joint materials used in the system and from the inadvertent introduction of nonpotable or toxic transfer fluids.

In an indirect connection system, a freeze-protected heat transfer fluid is circulated through a closed loop to a heat exchanger. The heat is then transferred indirectly to the potable water. The fluids in these systems are nonpotable and occasionally toxic. The type of heat exchanger used in these systems depends on the exact nature of the transfer fluid used.

The use of direct connection systems is typically limited to solar water-heating systems where the potable water is heated directly by the solar collector and circulated through the system and is suited for use only in areas where the water in the collectors is not subject to freezing.

Code Text: *Existing metallic water service piping used for electrical grounding shall not be replaced with nonmetallic pipe or tubing until other approved means of grounding is provided.*

Discussion and Commentary: In many buildings, equipment grounding is accomplished by bonding a grounding conductor to the interior metal water piping system. The continuity of this electrical ground is essential to maintaining the safety of the electrical system, and the replacement of a portion of that metal piping system with nonmetallic piping interrupts the continuity of that electrical grounding system and creates a potentially hazardous situation. When the replacement of any portion of an existing metal water piping system with nonmetallic piping is proposed, the method of electrical system grounding employed in the existing building must be approved by the code official.

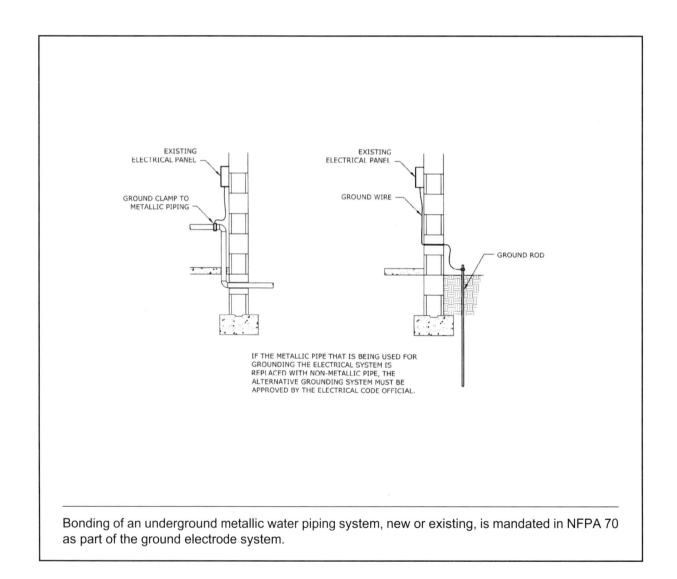

Bonding of an underground metallic water piping system, new or existing, is mandated in NFPA 70 as part of the ground electrode system.

Code Text: *Where a potable public water supply is not available, individual sources of potable water supply shall be utilized.*

Discussion and Commentary: An individual water supply is only permitted where a public system is unavailable. Individual sources of water supply may include: water wells, cisterns with treatment, reservoirs with treatment and other systems approved by the health department or authority having jurisdiction.

APPROVED: Acceptable to the code official or other authority having jurisdiction.

Code Text: *Dependent on geological and soil conditions and the amount of rainfall, individual water supplies are of the following types: drilled well, driven well, dug well, bored well, spring, stream or cistern. Surface bodies of water and land cisterns shall not be sources of individual water supply unless properly treated by approved means to prevent contamination.*

Discussion and Commentary: The sources for an individual water supply include wells, springs, cisterns, streams and surface water impoundments. Wells are probably the most common source of individual water supply used today and consist of a vertical shaft that extends down into the earth to a water-bearing strata. Wells may be bored, drilled, driven or dug. See Section 202 for the definitions of each well type and a cistern.

CISTERN. A small covered tank for storing water for a home or farm. Generally, this tank stores rainwater to be utilized for purposes other than in the potable water supply, and such tank is placed underground in most cases.

WELL

Bored. A well constructed by boring a hole in the ground with an auger and installing a casing.

Drilled. A well constructed by making a hole in the ground with a drilling machine of any type and installing casing and screen.

Driven. A well constructed by driving a pipe in the ground. The drive pipe is usually fitted with a well point and screen.

Dug. A well constructed by excavating a large-diameter shaft and installing a casing.

Surface water impoundments always have to be treated if used as a supply of potable water. This may include some form of chlorination or ozone treatment.

Code Text: *The combined capacity of the source and storage in an individual water supply system shall supply the fixtures with water at rates and pressures as required by* Chapter 6.

Discussion and Commentary: The amount and quality of water supplied to a building are critical to both the health of the occupants and the safe and efficient use of the plumbing fixtures and drainage system.

The design professional must determine the water demand rate for the building in accordance with the requirements of Section 604, Design of Building Water Distribution System, and in the same manner as if the water were supplied from a public source.

Topic: Disinfection of System

Reference: IPC 602.3.4

Category: Water Supply and Distribution

Subject: Water Required

Code Text: *After construction or major repair, the individual water supply system shall be purged of deleterious matter and disinfected in accordance with Section 610.*

Discussion and Commentary: A private water supply system must be disinfected after construction, major repair or alteration. The disinfection procedure must be as prescribed in Section 610, Disinfection of Potable Water System, for all potable water systems.

610.1 General. New or repaired potable water systems shall be purged of deleterious matter and disinfected prior to utilization. The method to be followed shall be that prescribed by the health authority or water purveyor having jurisdiction or, in the absence of a prescribed method, the procedure described in either AWWA C651 or AWWA C652, or as described in this section. This requirement shall apply to "on-site" or "in-plant" fabrication of a system or to a modular portion of a system.

1. The pipe system shall be flushed with clean, potable water until dirty water does not appear at the points of outlet.

2. The system or part thereof shall be filled with a water/chlorine solution containing at least 50 parts per million (50 mg/L) of chlorine, and the system or part thereof shall be valved off and allowed to stand for 24 hours; or the system or part thereof shall be filled with a water/chlorine solution containing at least 200 parts per million (200 mg/L) of chlorine and allowed to stand for 3 hours.

3. Following the required standing time, the system shall be flushed with clean potable water until the chlorine is purged from the system.

4. The procedure shall be repeated where shown by a bacteriological examination that contamination remains present in the system.

POTABLE WATER: Water free from impurities present in amounts sufficient to cause disease or harmful physiological effects and conforming to the bacteriological and chemical quality requirements of the Public Health Service Drinking Water Standards or the regulations of the public health authority having jurisdiction.

Topic: Pumps

Category: Water Supply and Distribution

Reference: IPC 602.3.5

Subject: Water Required

Code Text: *Pumps shall be rated for the transport of potable water. Pumps in an individual water supply system shall be constructed and installed so as to prevent contamination from entering a potable water supply through the pump units. Pumps shall be sealed to the well casing or covered with a water-tight seal. Pumps shall be designed to maintain a prime and installed such that ready access is provided to the pump parts of the entire assembly for repairs.*

Discussion and Commentary: A pump must be designed and installed to provide continuous service and is to be accessible to facilitate service, repair or replacement without requiring the removal or movement of any panel, door or similar obstruction and without the use of a portable ladder, step stool or similar device. The pump and its components that are in direct contact with the potable water must be rated for use in potable water systems.

The rating of a pump for use in potable water supply is generally determined by the pump's compliance with referenced standards related to materials used in potable water systems.

Topic: Pump Enclosure

Category: Water Supply and Distribution

Reference: IPC 602.3.5.1

Subject: Water Required

Code Text: *The pump room or enclosure around a well pump shall be drained and protected from freezing by heating or other approved means. Where pumps are installed in basements, such pumps shall be mounted on a block or shelf not less than 18 inches (457 mm) above the basement floor. Well pits shall be prohibited.*

Discussion and Commentary: A well pit, being below grade, has the potential for collecting rainwater, surface water or ground water and, subsequently, contaminating the well. The pump must be located so as to prevent the contamination of the well. When located in a basement, the pump must be elevated in the event that the basement becomes partially flooded. Pumps located outside a building must be properly protected from freezing during the winter months. Heat is typically provided by a small heating device controlled by a thermostat.

Well is a defined term in the IPC and consists of four types: bored, drilled, driven and dug.

| **Topic:** Separation from Building Sewer | **Category:** Water Supply and Distribution |
| **Reference:** IPC 603.2 | **Subject:** Water Service |

Code Text: *Water service pipe and the building sewer shall be separated by 5 feet (1524 mm) of undisturbed or compacted earth.*

Exceptions:
1. *The required separation distance shall not apply where the bottom of the water service pipe within 5 feet (1524 mm) of the sewer is a minimum of 12 inches (305 mm) above the top of the highest point of the sewer and the pipe materials conform to Table 702.3.*
2. *Water service pipe is permitted to be located in the same trench with a building sewer, provided such sewer is constructed of materials listed in Table 702.2.*
3. *The required separation distance shall not apply where a water service pipe crosses a sewer pipe provided the water service pipe is sleeved to at least 5 feet (1524 mm) horizontally from the sewer pipe centerline, on both sides of such crossing with pipe materials listed in Tables 605.3, 702.2 or 702.3.*

Discussion and Commentary: Contamination can occur when there is a leak in the building sewer located near the water service pipe. The soil then becomes contaminated around the water pipe, and, if the water service pipe has a subsequent failure, contamination of the potable water supply could occur.

The simplest means reducing the possibility of soil contamination is to install the building sewer and water service pipe in two trenches separated horizontally by undisturbed or compacted earth. Any contamination will tend to saturate the excavated soil before settling into the undisturbed, compacted earth. Exception 2 allows water and sewer pipes in the same trench.

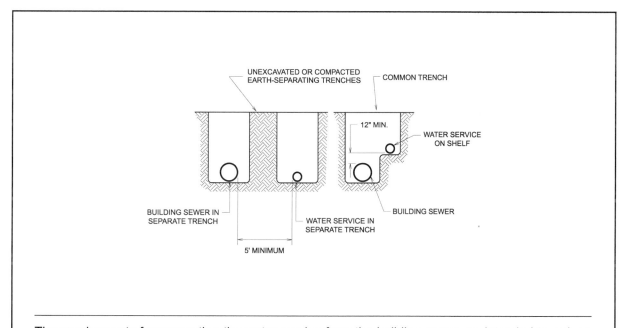

The requirements for separating the water service from the building sewer are intended to reduce the possibility of the sewer contaminating the potable water supply in the event of a failure of the water supply and building sewer pipes.

Code Text: *Hot water and cold water manifolds installed with gridded or parallel connected individual distribution lines to each fixture or fixture fitting shall be designed in accordance with Sections 604.10.1 through 604.10.3.*

Discussion and Commentary: Gridded and parallel water distribution systems are both permitted to be used by installers. This allows for more flexibility in using the most appropriate system based on the situation and need.

GRIDDED WATER DISTRIBUTION SYSTEM: A water distribution system where every water distribution pipe is interconnected so as to provide two or more paths to each fixture supply pipe.

Code Text: *The installation of a water service or water distribution pipe shall be prohibited in soil and ground water contaminated with solvents, fuels, organic compounds or other detrimental materials causing permeation, corrosion, degradation or structural failure of the piping material. Where detrimental conditions are suspected, a chemical analysis of the soil and ground water conditions shall be required to ascertain the acceptability of the water service or water distribution piping material for the specific installation. Where detrimental conditions exist, approved alternative materials or routing shall be required.*

Discussion and Commentary: When pipe is buried, the surrounding soil conditions might cause the pipe to degrade or corrode at an accelerated rate. The soil can be evaluated, or knowledge of the historical effects from the soil may be used to determine if additional requirements are necessary to protect the piping material. Most soils are corrosive to galvanized steel pipe. The pipe is either coated and wrapped with coal tar, or an elastomeric or epoxy coating is applied to the exterior of the pipe. Some soils affect copper tubing. Soil with cinders contains acid produced by combining water with sulfur compounds, which will attack unprotected copper pipe. Copper tubing in these instances must be coated with a protective layer.

Another example of degradation is thermoplastic pipe placed in soil containing heavy concentrations of hydrocarbons.

Code Text: *Water service pipe shall conform to NSF 61 and shall conform to one of the standards listed in Table 605.3. All water service pipe or tubing, installed underground and outside of the structure, shall have a minimum working pressure rating of 160 psi (1100 kPa) at 73.4°F (23°C). Where the water pressure exceeds 160 psi (1100 kPa), piping material shall have a minimum rated working pressure equal to the highest available pressure. Water service piping materials not third-party certified for water distribution shall terminate at or before the full open valve located at the entrance to the structure. All ductile iron water pipe shall be cement mortar lined in accordance with AWWA C104.*

Discussion and Commentary: Piping material that comes in contact with the potable water source is required to conform to the requirements of NSF 61. The intent is to control the potential adverse health effects produced by indirect additives, products and materials that come in contact with potable water. Plastic water pipe tested and labeled in accordance with NSF 14 as potable water pipe also conforms to NSF 61 insofar as NSF 14 makes reference to and requires compliance with NSF 61.

TABLE 605.3
WATER SERVICE PIPE

MATERIAL	STANDARD
Acrylonitrile butadiene styrene (ABS) plastic pipe	ASTM D 1527; ASTM D 2282
Asbestos-cement pipe	ASTM C 296
Brass pipe	ASTM B 43
Chlorinated polyvinyl chloride (CPVC) plastic pipe	ASTM D 2846; ASTM F 441; ASTM F 442; CSA B137.6
Copper or copper-alloy pipe	ASTM B 42; ASTM B 302
Copper or copper-alloy tubing (Type K, WK, L, WL, M or WM)	ASTM B 75; ASTM B 88; ASTM B 251; ASTM B 447
Cross-linked polyethylene (PEX) plastic tubing	ASTM F 876; ASTM F 877; CSA B137.5
Cross-linked polyethylene/aluminum/cross-linked polyethylene (PEX-AL-PEX) pipe	ASTM F 1281; CSA B137.10M
Cross-linked polyethylene/aluminum/high-density polyethylene (PEX-AL-HDPE)	ASTM F 1986
Ductile iron water pipe	AWWA C151; AWWA C115
Galvanized steel pipe	ASTM A 53
Polybutylene (PB) plastic pipe and tubing	ASTM D 2662; ASTM D 2666; ASTM D 3309; CSA B137.8M
Polyethylene (PE) plastic pipe	ASTM D 2239; CSA B137.1
Polyethylene (PE) plastic tubing	ASTM D 2737; CSA B137.1
Polyethylene/aluminum/polethylene (PE-AL-PE) pipe	ASTM F 1282; CSA B137.9
Polypropylene (PP) plastic pipe or tubing	ASTM F 2389; CSA B137.11
Polyvinyl chloride (PVC) plastic pipe	ASTM D 1785; ASTM D 2241; ASTM D 2672; CSA B137.3
Stainless steel pipe (Type 304/304L)	ASTM A 312; ASTM A 778
Stainless steel pipe (Type 316/316L)	ASTM A 312; ASTM A 778

Certain plastic piping materials, including chlorinated polyvinyl (CPVC) and cross-linked PEX, are permitted for both water service and water distribution. For all such materials, a transition or termination of the material would not be required.

Topic: Fittings	Category: Water Supply and Distribution
Reference: IPC 605.5	Subject: Materials, Joints and Connections

Code Text: *Pipe fittings shall be approved for installation with the piping material installed and shall conform to the respective pipe standards or one of the standards listed in Table 605.5. All pipe fittings utilized in water supply systems shall also conform to NSF 61. The fittings shall not have ledges, shoulders or reductions capable of retarding or obstructing flow in the piping. Ductile and gray iron pipe fittings shall be cement mortar lined in accordance with AWWA C104.*

Discussion and Commentary: Each fitting is designed to be installed in a particular system with a given material or combination of materials. Drainage pattern fittings must be installed in drainage systems. Vent fittings are limited to the venting system. Many fittings are intended to be used only for water distribution systems.

TABLE 605.5
PIPE FITTINGS

MATERIAL	STANDARD
Acrylonitrile butadiene styrene (ABS) plastic	ASTM D 2468
Cast-iron	ASME B16.4; ASME B16.12
Chlorinated polyvinyl chloride (CPVC) plastic	ASTM F 437; ASTM F 438; ASTM F 439; CSA B137.6
Copper or copper alloy	ASME B16.15; ASME B16.18; ASME B16.22; ASME B16.23; ASME B16.26; ASME B16.29
Cross-linked polyethylene/aluminum/high-density polyethylene (PEX-AL-HDPE)	ASTM F 1986
Fittings for cross-linked polyethylene (PEX) plastic tubing	ASTM F 877; ASTM F 1807; ASTM F 1960; ASTM F 2080; ASTM F 2159; CSA B137.5
Gray iron and ductile iron	AWWA C110; AWWA C153
Malleable iron	ASME B16.3
Metal (brass) insert fittings for Polyethylene/Aluminum/Polyethylene (PE-AL-PE) and Cross-linked Polyethylene/Aluminum/Polyethylene (PEX-AL-PEX)	ASTM F 1974
Polybutylene (PB) plastic	CSA B137.8
Polyethylene (PE) plastic	ASTM D 2609; CSA B137.1
Polypropylene (PP) plastic pipe or tubing	ASTM F 2389; CSA B137.11
Polyvinyl chloride (PVC) plastic	ASTM D 2464; ASTM D 2466; ASTM D 2467; CSA B137.2
Stainless steel (Type 304/304L)	ASTM A 312; ASTM A 778
Stainless steel (Type 316/316L)	ASTM A 312; ASTM A 778
Steel	ASME B16.9; ASME B16.11; ASME B16.28

Many pipe standards also include fittings. However, there are a number of other standards that strictly regulate pipe fittings. Pipe fittings are required to comply with the limitations set forth in NSF 61.

Topic: Polyethylene Plastic

Reference: IPC 605.20.4

Category: Water Supply and Distribution

Subject: Materials, Joints and Connections

Code Text: *Polyethylene pipe shall be cut square, with a cutter designed for plastic pipe. Except where joined by heat fusion, pipe ends shall be chamfered to remove sharp edges. Kinked pipe shall not be installed. The minimum pipe bending radius shall not be less than 30 pipe diameters, or the minimum coil radius, whichever is greater. Piping shall not be bent beyond straightening of the curvature of the coil. Bends shall not be permitted within 10 pipe diameters of any fitting or valve. Stiffener inserts installed with compression-type couplings and fittings shall not extend beyond the clamp or nut of the coupling or fitting.*

Discussion and Commentary: Any properly made pipe joint requires that the pipe ends be cut square for proper alignment, proper insertion depth and adequate surface area for joining. Because polyethylene plastic pipe and tubing can be kinked and stress-weakened, the minimum pipe bending radius is specified so as not to exceed the structural capacity of the material. The minimum bending radius must not be less than the radius of the pipe coil as it came from the manufacturer or 30 pipe diameters, whichever of these two criteria is the largest.

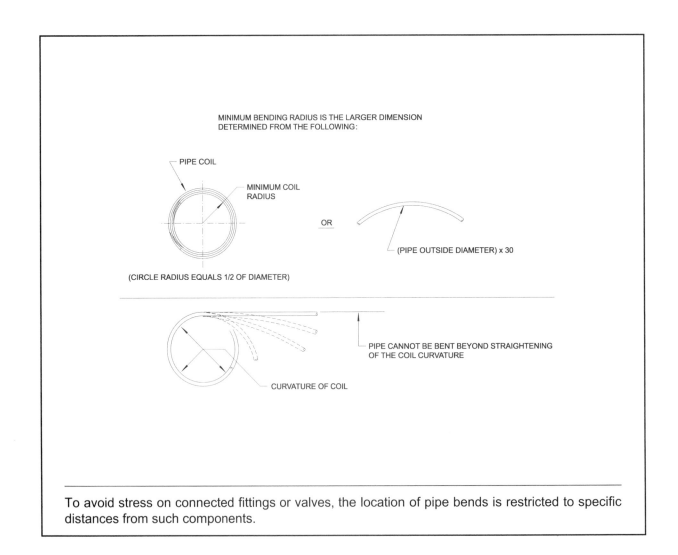

To avoid stress on connected fittings or valves, the location of pipe bends is restricted to specific distances from such components.

Code Text: *Joints between PP plastic pipe and fittings shall comply with Sections 605.21.1 or 605.21.2.*

Heat-fusion joints for polypropylene pipe and tubing joints shall be installed with socket-type heat-fused polypropylene fittings, butt-fusion polypropylene fittings or electrofusion polypropylene fittings. Joint surfaces shall be clean and free from moisture. The joint shall be undisturbed until cool. Joints shall be made in accordance with ASTM F 2389.

Mechanical and compression sleeve joints shall be installed in accordance with the manufacturer's instructions.

Discussion and Commentary: The requirements for mechanical and heat fusion joints are described separately to avoid misunderstanding and confusion. The details and methodology for mechanical and compression joints are governed by the manufacturer's installation instructions, whereas heat-fusion joints must meet certain code performance characteristics.

The joining methodologies described in Section 605.21 are applicable to polypropylene plastic piping (PP).

Code Text: *Mechanical joints on water pipe shall be made with an elastomeric seal conforming to ASTM D 3139. Mechanical joints shall not be installed in above-ground systems unless otherwise approved. Joints shall be installed in accordance with the manufacturer's instructions.*

Discussion and Commentary: A PVC plastic hub pipe with an elastomeric compression gasket is installed in underground sewer systems. The pipe must be inserted to the full depth of the hub to make a proper joint.

ASTM

ASTM International
100 Barr Harbor Drive
West Conshohocken, PA 19428-2959

Section	Material	Application	Standard	Title	Applies to IPC Sections
605.22.1	polyvinyl chloride (PVC) plastic pipe	Mechanical joints for underground water piping	ASTM D 3139	Specification for Joints for Plastic Pressure Pipes Using Flexible Elastomeric Seals	605.10.1, 605.22.1

The IPC defines four different joint types: expansion, flexible, mechanical and slip.

Quiz

Study Session 7
IPC Sections 601 – 605

1. A _____ is not permitted as a source for an individual water supply unless the water is treated to prevent contamination.

   a. spring   b. cistern

   c. dug well   d. drilled well

   Reference _____

2. The maximum allowable water pressure on the water distribution system is _____ .

   a. 25 psi   b. 40 psi

   c. 80 psi   d. 100 psi

   Reference _____

3. When the maximum allowable water pressure is exceeded, which of the following is required?

   a. flow control valve (vented)

   b. pressure reducing valve

   c. in-line pressure balancing valve

   d. pressure relief valve

   Reference _____

4. What is required to provide protection for the water distribution system against the effects of hydraulic shock?

    a. pressure reducing valve   b. flow control valve (vented)

    c. water hammer arrestor   d. pressure relief valve

Reference _____

5. Which of the following is the consequence of high velocity and the effect of a fast-acting mechanical valve on the water distribution system?

    a. thermal shock   b. water hammer

    c. hydraulic jump   d. galvanic action

Reference _____

6. What is the maximum allowable lead content of water supply pipe and fittings?

    a. .02 percent   b. .2 percent

    c. 2 percent   d. 8 percent

Reference _____

7. Which of the following is required to be installed with a pressure reducing valve?

    a. strainer   b. ball valve

    c. check valve   d. vacuum breaker

Reference _____

8. Which of the following pipe materials are approved for use in the water distribution system?

    a. PVC, ASTM D 1785   b. ductile iron, AWWA C151

    c. polyethylene, ASTM F 2239   d. polypropylene, ASTM F 2389

Reference _____

9. The minimum pressure rating for hot water distribution piping is _____ psi at _____ °F.

    a. 100, 180   b. 150, 210

    c. 160, 140   d. 120, 200

    Reference _____

10. Which of the following standards regulate/s drinking water system components?

    a. NSF 18   b. NSF 61

    c. ASSE 1016   d. ASSE 1017

    Reference _____

11. Which of the following transition fitting materials is approved for a connection between copper or copper-alloy and ferrous type metal piping?

    a. plastic   b. brass

    c. galvanized   d. stainless steel

    Reference _____

12. The pressure rating of Schedule 80 PVC or CPVC pipe is reduced by _____ percent when threaded.

    a. 10   b. 25

    c. 35   d. 50

    Reference _____

13. Where street main pressures are known to be variable, what is used as the design criteria for the water distribution system?

    a. mean available pressure

    b. maximum available pressure

    c. minimum available pressure

    d. supplemental pressure (pump assisted)

    Reference _____

14. The minimum pressure rating for water service piping is _____ psi at 73.4°F.

    a. 100   b. 150

    c. 160   d. 200

Reference _____

15. A parallel water distribution system has a supply branch from a manifold to a shower that is fifty feet in developed length. The available pressure at the meter is 60 PSI. What is the minimum required size of the fixture supply to the shower?

    a. $^1/_4$ inch   b. $^3/_8$ inch

    c. $^1/_2$ inch   d. $^5/_8$ inch

Reference _____

16. What is the maximum allowable lead composition in a solder classified as *lead free*?

    a. 0 percent   b. 0.2 percent

    c. 2 percent   d. 8 percent

Reference _____

17. The maximum water consumption flow rate for nonmetered public lavatories is _____ gpm.

    a. 0.25   b. 0.5

    c. 2   d. 2.2

Reference _____

18. The maximum water consumption flow rate for a blow-out type water closet is _____ gallons per flush.

    a. 1.6   b. 3.5

    c. 1.0   d. 2.2

Reference _____

19. Individual water supply system pumps installed in basements shall be mounted not less than _____ inches above the basement floor.

    a. 18   b. 16

    c. 12   d. 8

Reference _____

20. Water service pipe shall be not less than _____ inch(es) in diameter.

    a. $^3/_4$ b. $^1/_2$

    c. 1   d. $1^1/_4$

Reference _____

21. In general, water service pipe shall be separated from the building sewer by a minimum of _____ feet of undisturbed or compacted earth.

    a. 3   b. 4

    c. 5   d. 6

Reference _____

22. Water service pipe is permitted in the same trench as the building sewer, provided the sewer is constructed of _____.

    a. materials approved for the location

    b. cast-iron pipe

    c. PVC DWV Schedule 40 pipe

    d. materials approved for underground use within a building

Reference _____

23. Solvent cement joints of CPVC water piping require a primer if the pipe size is greater than _____ inches in diameter.

    a. $1^1/_4$ b. $1^1/_2$

    c. 2   d. $2^1/_2$

Reference _____

24. Where a primer is not used for cement joining CPVC water piping, the solvent cement used must be _____ in color.

      a. yellow   b. orange

      c. purple   d. blue

Reference _____

25. A flush valve urinal requires a minimum fixture supply pipe of _____ inch.

      a. $^3/_8$ b. $^1/_2$

      c. $^3/_4$ d. 1

Reference _____

2006 IPC Sections 606 – 613
Water Supply and Distribution II

OBJECTIVE: To develop an understanding of the code provisions related to potable water supply and its protection against contamination. To develop an understanding of the provisions that ensure the supply of adequate cold and hot water to each fixture.

REFERENCE: Sections 606 through 613, 2006 *International Plumbing Code*

KEY POINTS:
- When piping is used for water, what limitations apply to the solder used?
- Are mechanical joints allowed above ground?
- What types of joints are permitted between different materials?
- What types of joints are required between plastic pipe and cast-iron hub pipe?
- What three types of fittings are allowed for joints between stainless steel and other piping materials?
- Where are full-open valves required?
- Where are shutoff valves required?
- Access shall be provided to which valves?
- Which valves are required to be identified?
- At what point is a booster required, and what type of booster can be used?
- What is the minimum height of the air gap above the overflow?
- What is used to allow emptying of the tank, and where is it to be located?
- Where is the vacuum relief valve located?
- When is a water system to be tested?
- What type of structure requires hot water?
- At what length of hot water piping does the code require a method of maintaining temperature?
- When is insulation required on water supply piping?
- What requirements apply to shutoff valves, and when should they be used?
- How is the hot or tempered water return from a thermostatic mixing valve to be connected?
- What is a byproduct of thermal expansion?

KEY POINTS:
(Cont'd)

- Under what circumstances is a device for controlling pressure to be installed?
- Which is the correct side for the installation of hot water to a fixture faucet?
- What are the three types of devices that may be used for hospital fixtures to prevent backflow?
- When is identification of a water system required?
- What is required to be included in the pipe identifications?
- What type of water is not permitted to be returned to the potable water system?
- Does the code allow the reuse of pipe for a potable water system?
- Where is the barometric loop located, and to what height is it extended?
- What limitations apply to where pressure-type vacuum breakers can be installed?
- Is access required to backflow preventers?
- What is required as protection between the water supply connection and a carbonated beverage dispenser?

Topic: Location of Shutoff Valves

Category: Water Supply and Distribution

Reference: IPC 606.2

Subject: Water Distribution System Installation

Code Text: *Shutoff valves shall be installed in the following locations:*

1. *On the fixture supply to each plumbing fixture other than bathtubs and showers in one- and two-family residential occupancies, and other than in individual sleeping units that are provided with unit shutoff valves in hotels, motels, boarding houses and similar occupancies.*
2. *On the water supply pipe to each sillcock.*
3. *On the water supply pipe to each appliance or mechanical equipment.*

Discussion and Commentary: A shutoff valve, unlike a full-open valve, has no requirements for the cross-sectional area of flow. Therefore, there is a greater pressure drop through a shutoff valve when compared to a full-open valve. Shutoff valves are commonly referred to as *stops* and include globe valves and straight and angle stops.

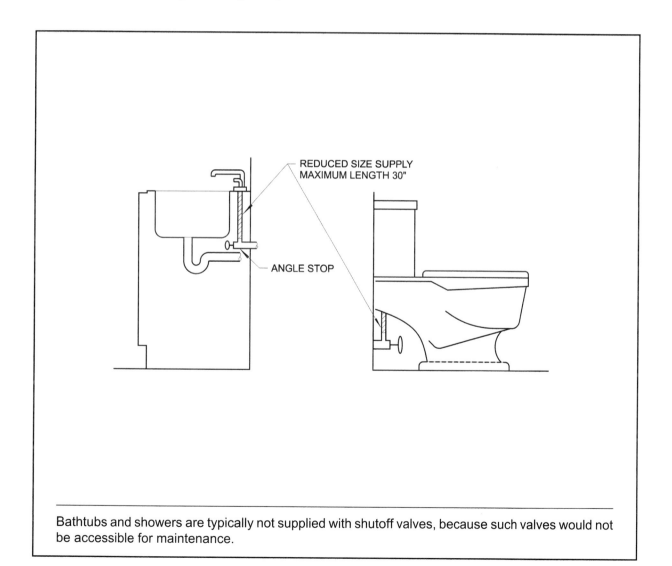

REDUCED SIZE SUPPLY
MAXIMUM LENGTH 30"

ANGLE STOP

Bathtubs and showers are typically not supplied with shutoff valves, because such valves would not be accessible for maintenance.

Code Text: *Access shall be provided to all full-open valves and shutoff valves.*

Discussion and Commentary: Valves of full-open type and shutoff valves must be accessible for the purposes of repair and maintenance. This requirement is not based on whether such valves are required by the code. All such valves, required by code or installed by choice, must be accessible.

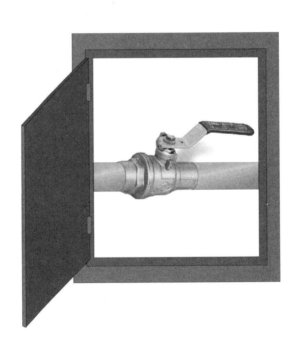

The code permits an access cover such as a panel, door or similar obstruction that must be moved or removed to gain access to the valve.

Topic: Where Required
Reference: IPC 607.1

Category: Water Supply and Distribution
Subject: Hot Water Supply System

Code Text: *In residential occupancies, hot water shall be supplied to all plumbing fixtures and equipment utilized for bathing, washing, culinary purposes, cleansing, laundry or building maintenance. In nonresidential occupancies, hot water shall be supplied for culinary purposes, cleansing, laundry or building maintenance purposes. In nonresidential occupancies, hot water or tempered water shall be supplied for bathing and washing purposes. Tempered water shall be supplied through a water temperature limiting device that conforms to ASSE 1070 and shall limit the tempered water to a maximum of 110°F (43°C). This provision shall not supersede the requirement for protective shower valves in accordance with Section 424.3.*

Discussion and Commentary: This section establishes the requirement for hot water and regulates the installation of hot water supply systems. Hot water is necessary primarily for convenience and the comfort of building occupants. In certain occupancies, tempered water [water ranging in temperature from 85°F to 110°F (29°C to 43°C) (see the definition of *tempered water* in Section 202) may be supplied in lieu of hot water. For example, in a restaurant, tempered water could be supplied to a hand-washing sink instead of hot water. Tempered water must be supplied to all accessible hand-washing facilities, regardless of the occupancy.

The code defines hot water as water at a temperature of 110°F (43°C) or greater.

Topic: Flow of Hot Water to Fixtures

Category: Water Supply and Distribution

Reference: IPC 607.4

Subject: Hot Water Supply System

Code Text: *Fixture fittings, faucets and diverters shall be installed and adjusted so that the flow of hot water from the fittings corresponds to the left-hand side of the fixture fitting. Exception: Shower and tub/shower mixing valves conforming to ASSE 1016 or CSA B125, where the flow of hot water corresponds to the markings on the device.*

Discussion and Commentary: One of the oldest expressions in plumbing is: "Hot on the left, cold on the right." The code mandates that hot corresponds to the left side of the fixture fitting for safety reasons. It has become an accepted practice to equate the left side of the faucet with hot water. The intent is to protect an individual from potential scalding when turning on what is believed to be cold water is actually hot water.

FAUCET: A valve end of a water pipe through which water is drawn from or held within the pipe.

Code Text: *A potable water supply system shall be designed, installed and maintained in such a manner so as to prevent contamination from nonpotable liquids, solids or gases being introduced into the potable water supply through cross-connections or any other piping connections to the system. Backflow preventer applications shall conform to Table 608.1, except as specifically stated in Sections 608.2 through 608.16.10.*

Discussion and Commentary: The most important aspect of a plumbing code is the protection of potable water systems. Documented local and widespread occurrences of sickness and disease have occurred because of inadequate safeguarding of the water supply.

TABLE 608.1
APPLICATION OF BACKFLOW PREVENTERS

DEVICE	DEGREE OF HAZARD[a]	APPLICATION[b]	APPLICABLE STANDARDS
Air gap	High or low hazard	Backsiphonage or backpressure	ASME A112.1.2
Air gap fittings for use with plumbing fixtures, appliances and appurtenances	High or low hazard	Backsiphonage or backpressure	ASME A112.1.3
Antisiphon-type fill valves for gravity water closet flush tanks	High hazard	Backsiphonage only	ASSE 1002, CSA B125
Backflow preventer for carbonated beverage machines	Low hazard	Backpressure or backsiphonage Sizes $1/4''$ - $3/8''$	ASSE 1022, CSA B64.3.1
Backflow preventer with intermediate atmospheric vents	Low hazard	Backpressure or backsiphonage Sizes $1/4''$ - $3/4''$	ASSE 1012, CSA B64.3
Barometric loop	High or low hazard	Backsiphonage only	(See Section 608.13.4)
Double check backflow prevention assembly and double check fire protection backflow prevention assembly	Low hazard	Backpressure or backsiphonage Sizes $3/8''$ - 16"	ASSE 1015, AWWA C510, CSA B64.5, CSA B64.5.1
Double check detector fire protection backflow prevention assemblies	Low hazard	Backpressure or backsiphonage (Fire sprinkler systems) Sizes 2" - 16"	ASSE 1048
Dual-check-valve-type backflow preventer	Low hazard	Backpressure or backsiphonage Sizes $1/4''$ - 1"	ASSE 1024, CSA B64.6
Hose connection backflow preventer	High or low hazard	Low head backpressure, rated working pressure, backpressure or backsiphonage Sizes $1/2''$-1"	ASSE 1052, CSA B64.2.1.1
Hose connection vacuum breaker	High or low hazard	Low head backpressure or backsiphonage Sizes $1/2''$, $3/4''$, 1"	ASSE 1011, CSA B64.2, CSA B64.2.1
Laboratory faucet backflow preventer	High or low hazard	Low head backpressure and backsiphonage	ASSE 1035, CSA B64.7
Pipe-applied atmospheric-type vacuum breaker	High or low hazard	Backsiphonage only Sizes $1/4''$ - 4"	ASSE 1001, CSA B64.1.1
Pressure vacuum breaker assembly	High or low hazard	Backsiphonage only Sizes $1/2''$ - 2"	ASSE 1020, CSA B64.1.2
Reduced pressure principle backflow preventer and reduced pressure principle fire protection backflow preventer	High or low hazard	Backpressure or backsiphonage Sizes $3/8''$ - 16"	ASSE 1013, AWWA C511, CSA B64.4, CSA B64.4.1
Reduced pressure detector fire protection backflow prevention assemblies	High or low hazard	Backsiphonage or backpressure (Fire sprinkler systems)	ASSE 1047
Spillproof vacuum breaker	High or low hazard	Backsiphonage only Sizes $1/4''$-2"	ASSE 1056
Vacuum breaker wall hydrants, frost-resistant, automatic draining type	High or low hazard	Low head backpressure or backsiphonage Sizes $1/4''$, 1"	ASSE 1019, CSA B64.2.2

For SI: 1 inch = 25.4 mm.
a. Low hazard–See Pollution (Section 202).
   High hazard–See Contamination (Section 202).
b. See Backpressure (Section 202).
   See Backpressure, low head (Section 202).
   See Backsiphonage (Section 202).

It is imperative that the potable water supply be maintained in a safe-for-drinking condition at all times and at all outlets.

Code Text: *The supply lines and fittings for every plumbing fixture shall be installed so as to prevent backflow. Plumbing fixture fittings shall provide backflow protection in accordance with ASME A112.18.1.*

Discussion and Commentary: To prevent the potable water supply system from being contaminated, fixtures are required to be installed in a manner that will prevent backflow. Many fixtures, such as water closets, come equipped with an integral backflow prevention device. For fixtures that do not have an integral backflow preventions device, an approved backflow preventer must be installed. The American Society of Mechanical Engineers (ASME) Standard A112.18.1 regulates backflow prevention devices.

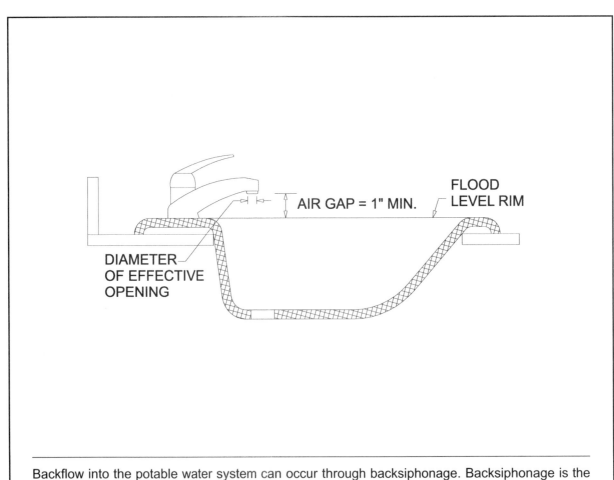

Backflow into the potable water system can occur through backsiphonage. Backsiphonage is the backflow of potentially contaminated water into the potable water system as a result of the pressure in the potable water system falling below atmospheric pressure of the plumbing fixtures, pools, tanks or vats connected to the potable water distribution piping.

Code Text: *Reduced pressure principle backflow preventers shall conform to ASSE 1013, AWWA C511, CSA B64.4 or CSA B64.4.1. Reduced pressure detector assembly backflow preventers shall conform to ASSE 1047. These devices shall be permitted to be installed where subject to continuous pressure conditions. The relief opening shall discharge by air gap and shall be prevented from being submerged.*

Discussion and Commentary: A reduced pressure principle backflow preventer is considered to be the most reliable mechanical method for preventing backflow. These devices consist of dual independently acting, spring-loaded check valves that are separated by a chamber or *zone* equipped with a relief valve. The pressure downstream of the device and in the central chamber between the check valves is maintained at a minimum of 2 pounds psi (13.8 kPa) less than the potable water supply pressure at the device inlet, hence the name *reduced pressure principle*. The relief valve located in the central chamber is held closed by the pressure differential between the inlet supply pressure and the central chamber pressure.

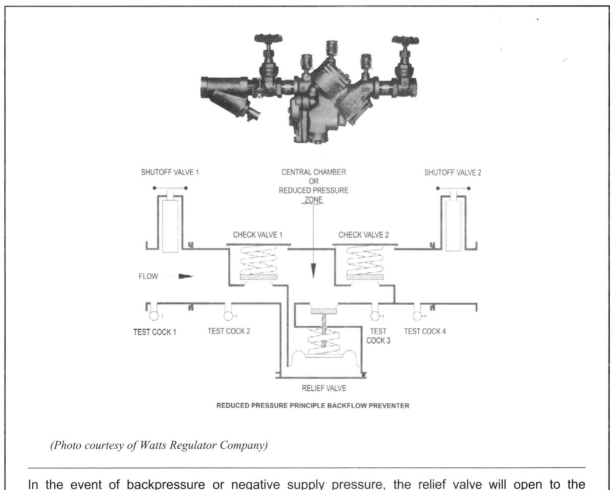

REDUCED PRESSURE PRINCIPLE BACKFLOW PREVENTER

(Photo courtesy of Watts Regulator Company)

In the event of backpressure or negative supply pressure, the relief valve will open to the atmosphere and drain any backflow that has leaked through the check valve. The relief vent will also allow air to enter to prevent any siphonage.

Code Text: *Barometric loops shall precede the point of connection and shall extend vertically to a height of 35 feet (10 668 mm). A barometric loop shall only be utilized as an atmospheric-type or pressure-type vacuum breaker.*

Discussion and Commentary: The code considers a barometric loop to be equivalent to a vacuum breaker and would allow it to be installed in any location where a vacuum breaker would otherwise be required.

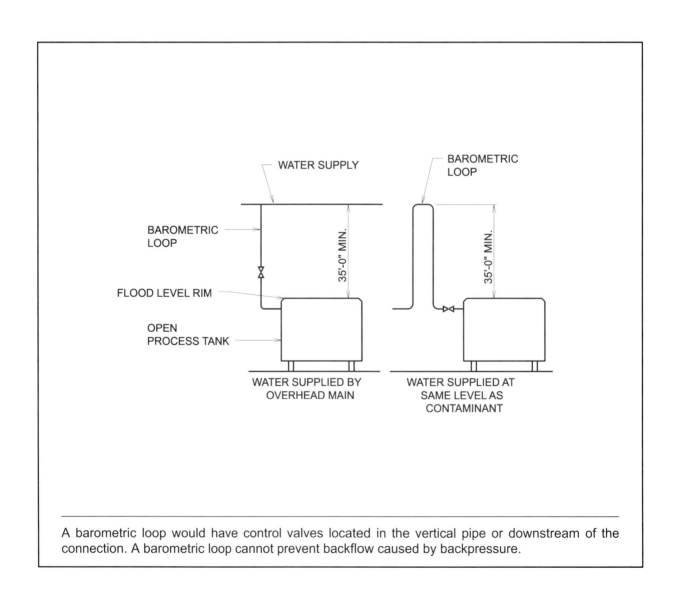

A barometric loop would have control valves located in the vertical pipe or downstream of the connection. A barometric loop cannot prevent backflow caused by backpressure.

Code Text: *Pipe-applied atmospheric-type vacuum breakers shall conform to ASSE 1001 or CSA-B64.1.1. Hose-connection vacuum breakers shall conform to ASSE 1011, ASSE 1019, ASSE 1035, ASSE 1052, CSA-B64.2, CSA-B64.2.1, CSA B64.2.1.1, CSA B64.2.2 or CSA B64.7. These devices shall operate under normal atmospheric pressure when the critical level is installed at the required height.*

Discussion and Commentary: An atmospheric vacuum breaker is designed to prevent siphonic action from occurring downstream of the device. When siphon action or a vacuum is applied to the water supply, the vacuum breaker opens to the supply inlet of the vacuum breaker, and the disc float drops over the opening. Air enters the atmospheric port, opening allowing the remaining water in the piping downstream from the vacuum breaker to drain.

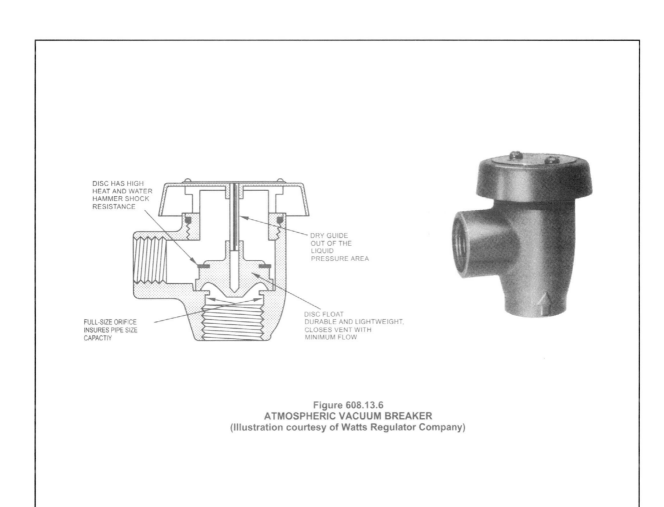

Figure 608.13.6
ATMOSPHERIC VACUUM BREAKER
(Illustration courtesy of Watts Regulator Company)

Valves must not be installed downstream of this device, as this would subject the device to supply pressure, thereby rendering it inoperative.

Code Text: *Double check- valve assemblies shall conform to ASSE 1015, CSA B64.5, CSA B64.5.1 or AWWA C510. Double-detector check-valve assemblies shall conform to ASSE 1048. These devices shall be capable of operating under continuous pressure conditions.*

Discussion and Commentary: These devices are designed for low-hazard applications subject to backpressure and backsiphonage applications. The devices consist of two independent spring-loaded check valves in series. Test cocks are provided to permit testing of the devices.

Double Check-valve Assembly Backflow Preventer

Photo courtesy of Watts Regulator Company

Note that these devices must not be confused with dual check-valve devices or two single check-valves placed in a series.

Code Text: *The potable water supply to automatic fire sprinkler and standpipe systems shall be protected against backflow by a double check-valve assembly or a reduced pressure principle backflow preventer.*

Exceptions:
1. *Where systems are installed as a portion of the water distribution system in accordance with the requirements of this code and are not provided with a fire department connection, isolation of the water supply system shall not be required.*
2. *Isolation of the water distribution system is not required for deluge, preaction or dry pipe systems.*

Discussion and Commentary: A double check-valve assembly is the minimum form of backflow prevention required between the potable water supply and an automatic fire sprinkler system or standpipe system. Protection by a double check-valve assembly is permitted only where the sprinkler or standpipe system is filled from a potable water source.

If antifreeze or other chemicals are added to a sprinkler or standpipe system, or if a nonpotable hazardous secondary supply system is involved, the potable water supply must be protected with a reduced pressure principle backflow preventer.

Code Text: *Where systems under continuous pressure contain chemical additives or antifreeze, or where systems are connected to a nonpotable secondary water supply, the potable water supply shall be protected against backflow by a reduced pressure principle backflow preventer. Where chemical additives or antifreeze are added to only a portion of an automatic fire sprinkler or standpipe system, the reduced pressure principle backflow preventer shall be permitted to be located so as to isolate that portion of the system. Where systems are not under continuous pressure, the potable water supply shall be protected against backflow by an air gap or a pipe applied atmospheric vacuum breaker conforming to ASSE 1001 or CAN/CSA B64.1.1.*

Discussion and Commentary: A nonpotable secondary water supply could include above- or below-ground tanks, private wells, ponds, reservoirs and lakes. In the event of loss or inadequacy of the primary potable supply, the secondary nonpotable water source is pumped into the system, thereby contaminating the potable supply in the event backflow occurs. Fire suppression systems that are not under continuous pressure are classified as indirect cross connections. When subject to only backsiphonage, these systems may be protected by an atmospheric-type vacuum breaker conforming to ASSE 1001 or CSA B64.1.1. These devices protect the potable water supply against pollutants or contaminants that enter the system because of backsiphonage through the outlet.

POLLUTION: An impairment of the quality of the potable water to a degree that does not create a hazard to the public health but that does adversely and unreasonably affect the aesthetic qualities of such potable water for domestic use.

| **Topic:** Coffee Beverage Dispensers | **Category:** Water Supply and Distribution |
| **Reference:** IPC 608.16.10 | **Subject:** Potable Water Supply Protection |

Code Text: *The water supply connection to coffee machines and noncarbonated beverage dispensers shall be protected against backflow by a backflow preventer conforming to ASSE 1022 or by an air gap.*

Discussion and Commentary: Coffee machines, similar to beverage dispensing machines, have the potential for creating cross connection and as such should be protected by a backflow preventer or an integral air gap just as beverage dispensing machines are. A coffee machine functions much like a small boiler.

An ASSE 1022 device is similar to the protection accepted on a boiler without chemical additives.

Topic: Water-tight Casings	**Category:** Water Supply and Distribution
Reference: IPC 608.17.4	**Subject:** Individual Water Supply Protection

Code Text: *Each well shall be provided with a water-tight casing to a minimum distance of 10 feet (3048 mm) below the ground surface. All casings shall extend at least 6 inches (152 mm) above the well platform. The casing shall be large enough to permit installation of a separate drop pipe. Casings shall be sealed at the bottom in an impermeable stratum or extend several feet into the water-bearing stratum.*

Discussion and Commentary: Any contaminant can enter the well with relative ease if the well does not have a proper casing. High-water tables are more susceptible to contamination because of their close proximity to the ground surface.

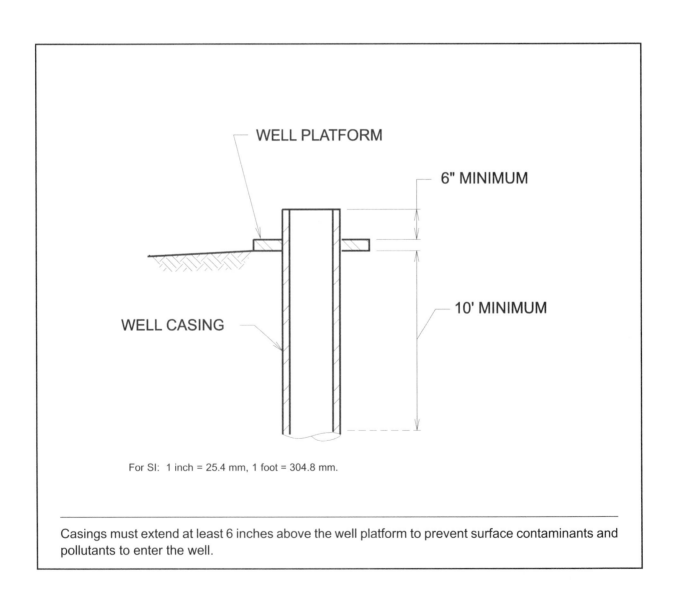

WELL PLATFORM

6" MINIMUM

10' MINIMUM

WELL CASING

For SI: 1 inch = 25.4 mm, 1 foot = 304.8 mm.

Casings must extend at least 6 inches above the well platform to prevent surface contaminants and pollutants to enter the well.

Topic: Cover

Reference: IPC 608.17.7

Category: Water Supply and Distribution

Subject: Individual Water Supply Protection

Code Text: *Every potable water well shall be equipped with an overlapping water-tight cover at the top of the well casing or pipe sleeve such that contaminated water or other substances are prevented from entering the well through the annular opening at the top of the well casing, wall or pipe sleeve. Covers shall extend downward at least 2 inches (51 mm) over the outside of the well casing or wall. A dug well cover shall be provided with a pipe sleeve permitting the withdrawal of the pump suction pipe, cylinder or jet body without disturbing the cover. Where pump sections or discharge pipes enter or leave a well through the side of the casing, the circle of contact shall be water tight.*

Discussion and Commentary: A cover is important to protect the well water from contamination from an outside source. It must fit firmly to the casing with cracks or openings filled tightly to create a watertight seal.

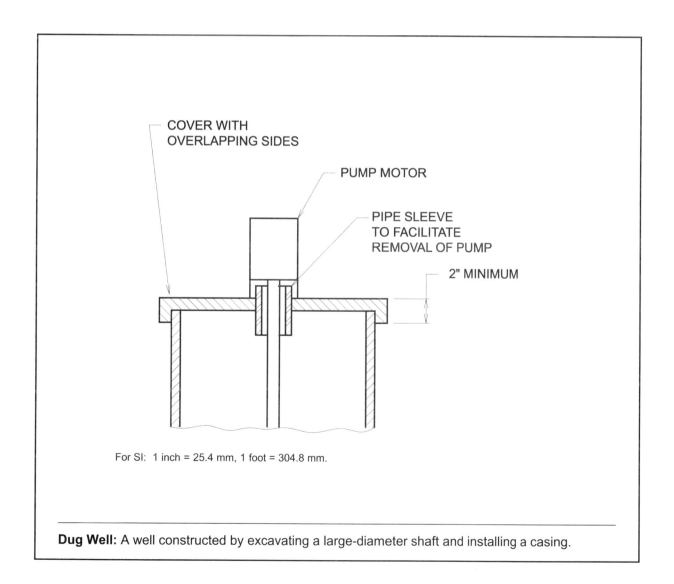

For SI: 1 inch = 25.4 mm, 1 foot = 304.8 mm.

Dug Well: A well constructed by excavating a large-diameter shaft and installing a casing.

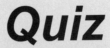

Study Session 8
2006 IPC Sections 606 – 613

1. All of the following locations in the water distribution system require the installation of a shutoff valve, *except* _____ .

    a. the fixtures supplies to a kitchen sink

    b. the supply connection to a sillcock

    c. the supply to mechanical equipment

    d. the supply to a water heater

    Reference _____

2. Which valves in the distribution system are required to be identified?

    a. all valves

    b. service valves only

    c. sillcock valves, pressure reducing valves and backflow prevention valves

    d. service valves, hose bibb valves and those remote from the fixtures served

    Reference _____

3. A temperature limiting device that conforms to ASSE 1070 performs which of the following functions?

    a. prevents pressure fluctuations during thermal expansion

    b. prevents pressure fluctuations that result in temperature shock at 140°F

    c. regulates the fixture water supply temperature to a maximum of 110°F

    d. regulates the fixture water supply temperature to a maximum of 120°F

Reference _____

4. A water supply tank that has a filling capacity of 175 GPM is required to have an overflow that is a minimum diameter of _____ inches.

    a. $2^1/_2$ b. 3

    c. 4   d. 6

Reference _____

5. A pressure spike in the water distribution system occurring during thermal expansion is caused by which of the following installations?

    a. a thermal expansion tank at the water heater

    b. a vacuum relief valve at the water heater

    c. a check-valve in the water service

    d. an elevated water supply source

Reference _____

6. Which of the following devices is not approved for isolating a high-hazard cross-connection?

    a. spillproof vacuum breaker

    b. double-check backflow preventer

    c. hose connection vacuum breaker

    d. reduced pressure principle backflow preventer

Reference _____

7. Which of the following connections to a potable water supply is required to be isolated from the water distribution system by a reduced pressure principle backflow preventer?

    a. carbonated beverage dispenser

    b. air gap connection to a livestock watering tank

    c. double-wall heat exchanger with toxic transfer fluid

    d. fire sprinkler system with a nonpotable secondary supply

    Reference _____

8. Which of these fixtures are not required to have the flow of hot water oriented to the left-hand side of the fixture fitting?

    a. laundry tub faucet   b. kitchen sink faucet.

    c. whirlpool filler valve   d. ASSE 1016, mixing valve

    Reference _____

9. Which of the following equipment does not require backflow protection on the water supply connection?

    a. ice machine

    b. coffee machine

    c. carbonated beverage dispenser

    d. noncarbonated beverage dispenser

    Reference _____

10. An irrigation system that has been designed for application of chemicals using a pump assisted chemical hopper must be isolated from the potable water supply by which of the following backflow preventers?

    a. barometric loop

    b. atmospheric vacuum breaker

    c. pressure vacuum breaker assembly

    d. reduced pressure principle backflow preventer

    Reference _____

11. What is the proper frequency for labeling of a nonpotable piping system in a building?

    a. at the entrance to the building and at the point of use

    b. wherever the piping is accessible for maintenance or repair

    c. where the nonpotable system is visible alongside the potable water system

    d. 25-foot intervals, and at each penetration of a wall, floor or roof assembly

Reference _____

12. The minimum air gap required for a 1-inch whirlpool tub filler spout that is located at the center of the tub and away from sidewalls is _____ .

    a. 1 inch   b. 1.5 inches

    c. 2 inches   d. 3 inches

Reference _____

13. Heat exchangers utilizing a transfer fluid with a Gosselin rating of one shall be separated from the potable water by _____ .

    a. single-wall construction

    b. double-wall construction

    c. triple-wall construction

    d. a double check-valve assembly

Reference _____

14. Where the water table is at 20 feet, the minimum length of well casing required for a bored well is _____ feet.

    a. 10   b. 11

    c. 20   d. 21

Reference _____

15. The minimum concentration of chlorine required for disinfection of a potable water supply system is _____ parts per million for 24 hours.

    a. 25   b. 50

    c. 100   d. 200

Reference _____

16. A vacuum breaker serving a bedpan washer is required to be located a minimum of _____ feet above the floor.

    a. 0.5   b. 2

    c. 5   d. 6

Reference _____

17. Which of the following is required to prevent the collapse of an elevated pressure tank in a water pressure booster system?

    a. check-valve   b. vacuum breaker

    c. vacuum relief valve   d. pressure reducing valve

Reference _____

18. What is the minimum height required for a barometric loop?

    a. 34 feet   b. 35 feet

    c. 43.3 feet   d. 100 feet

Reference _____

19. Which of the following measurements represents the physical protection of the water supply system against backflow or backsiphonage by separation with an air gap?

    a. the lowest end of the supply outlet to inlet of the fixture drain

    b. the lowest end of the supply outlet to the crown weir of a drainage trap

    c. the lowest end of the supply outlet to the flood level rim of the receptacle

    d. the lowest end of the supply to a point in the receptacle, above the trap seal

Reference _____

20. Where both potable and nonpotable water systems are installed in the same building, _____ is mandatory.

    a. identification of only the potable system

    b. identification of only the nonpotable system

    c. identification of both the nonpotable and potable systems

    d. identification of the nonpotable system when interconnected with the potable system

Reference _____

21. A backflow preventer that is not affected by carbon dioxide gas and is designed to isolate carbon dioxide sources from backflow into the potable water system conforms to which of the following standards?

    a. ASSE 1012   b. ASSE 1013

    c. ASSE 1022   d. ASSE 1024

Reference _____

22. Where is recirculation of a hot water supply piping system required?

    a. in all public access buildings

    b. in food processing and manufacturing facilities

    c. where the developed length from the heating source to the farthest fixture exceeds 100 feet

    d. where the developed length from the heating source to the farthest high consumption fixture exceeds 100 feet

Reference _____

23. All of the following could be an approved source for an individual water supply *except* _____ .

    a. water from a cistern, with an approved filtration and disinfection system

    b. water from a man-made pond, with an approved filtration system and a treatment system conforming to NSF 53

    c. water from a rainwater collection reservoir that passes through a treatment system conforming to NSF 14

    d. water from a lake that also supplies water for a livestock feedlot operation, when properly treated and disinfected

Reference _____

24. Which of the following is not within the scope of Chapter 6?

    a.   the installation of individual water wells

    b.   the installation of a solar swimming pool heater

    c.   the connection of a reverse osmosis water treatment system

    d.   the water supply connection to portable cleaning equipment

Reference _____

25. If a water purveyor were unable to provide a water supply at flow pressures sufficient to supply the minimum required pressures at the fixtures, all of the following would be a suitable resolution, *except* _____ .

    a.   oversizing the water distribution piping system

    b.   a hydropneumatic pressure booster system

    c.   a water pressure booster pump system

    d.   an elevated supplemental water supply

Reference _____

2006 IPC Sections 701 – 707
Sanitary Drainage I

OBJECTIVE: To develop an understanding of the overall code provisions that regulate the materials, design and installation of sanitary drainage. To develop an understanding of the specific provisions of the code that apply to joints and fittings in the sanitary drainage system.

REFERENCE: Sections 701 – 707, 2006 *International Plumbing Code*

KEY POINTS:
- Under what circumstances would a building not have to connect to a public sewer?
- When is a building required to have a separate sewer connection?
- Is steam exhaust, blowoff or drip pipe permitted to be directly connected to the building drainage system?
- What provisions are made to protect the sanitary drainage system from a chemical waste system?
- What distance between the sewer pipe and the water service is considered acceptable for underground installations?
- Whenever filled or unstable ground is present, what materials are permitted for piping in the ground, and which standards apply?
- Are sanitary and storm building sewers or drains permitted to be laid in one trench?
- What is necessary to prove an existing system can be connected to a new system?
- What type of alignment is required for horizontal drainage piping?
- Is it permitted to reduce the size of drainage pipe?
- What schedule of pipe is permitted to be threaded?
- Which section applies to joints between cast-iron pipe or fittings?
- Are solvent cement joints permitted below ground?
- What type of material is applied, and which standard applies to any filler metal used?
- What type of connections are required for glass-to-glass connections?
- What type of joint is to be used to join plastic pipe to other piping material?
- Which section provides the requirements for slip joints in drainage systems?
- Are there limitations for the use of a double sanitary tee?

Topic: Separate Sewer Connection

Category: Sanitary Drainage

Reference: IPC 701.3

Subject: General Provisions

Code Text: *Every building having plumbing fixtures installed and intended for human habitation, occupancy or use on premises abutting on a street, alley or easement in which there is a public sewer shall have a separate connection with the sewer. Where located on the same lot, multiple buildings shall not be prohibited from connecting to a common building sewer that connects to the public sewer.*

Discussion and Commentary: This requirement prohibits the combining of sewers serving different buildings prior to connection to the public sewer. The only exception is where the sewers to be combined are serving buildings on the same lot or parcel of land. This section does not prohibit the use of adjoining properties that have been included in a dedicated easement approved by the administrative authority. The common building sewer is an extension of the public sewer and under control of the public authority.

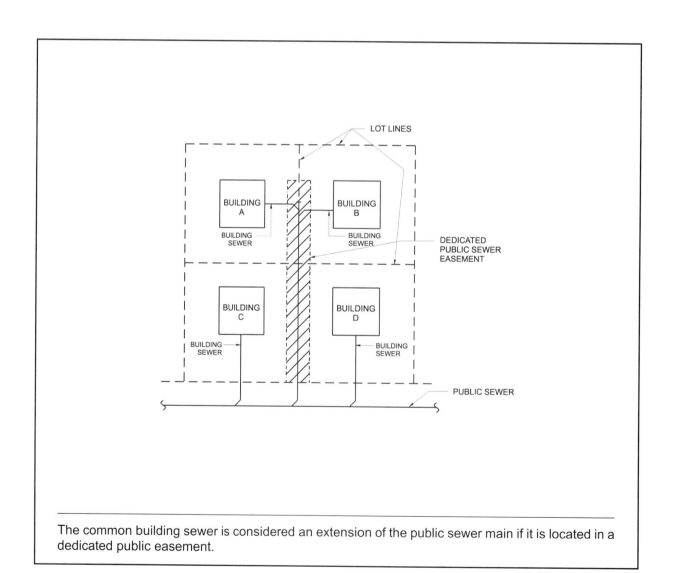

The common building sewer is considered an extension of the public sewer main if it is located in a dedicated public easement.

Topic: Chemical Waste System	**Category:** Sanitary Drainage
Reference: IPC 702.5	**Subject:** Materials

Code Text: *A chemical waste system shall be completely separated from the sanitary drainage system. The chemical waste shall be treated in accordance with Section 803.2 before discharging to the sanitary drainage system. Separate drainage systems for chemical wastes and vent pipes shall be of an approved material that is resistant to corrosion and degradation for the concentrations of chemicals involved.*

Discussion and Commentary: There is no universal material available to resist all chemical action. Each pipe has qualities that make it resistant to certain chemical waste systems. An individual analysis of compatibility is required for the installation of a chemical waste and vent system. Piping material such as borosilicate glass and polyolefin plastic is often used for special waste applications Acid wastes in the vapor state are more corrosive than in the liquid state; therefore, it is important to provide material that resists chemical action for the vent piping as well as for the drainage piping.

803.2 Neutralizing device required for corrosive wastes. Corrosive liquids, spent acids or other harmful chemicals that destroy or injure a drain, sewer, soil or waste pipe, or create noxious or toxic fumes or interfere with sewage treatment processes shall not be discharged into the plumbing system without being thoroughly diluted, neutralized or treated by passing through an approved dilution or neutralizing device. Such devices shall be automatically provided with a sufficient supply of diluting water or neutralizing medium so as to make the contents noninjurious before discharge into the drainage system. The nature of the corrosive or harmful waste and the method of its treatment or dilution shall be approved prior to installation.

A measure of the degree of acidity or alkalinity of a liquid is its pH value, which defines the concentration of free hydrogen ions on a scale of 0 to 14. A pH value of 7 is considered neutral.

Topic: Slope of Horizontal Drainage Piping

Category: Sanitary Drainage

Reference: IPC 704.1

Subject: Drainage Piping Installation

Code Text: *Horizontal drainage piping shall be installed in uniform alignment at uniform slopes. The minimum slope of a horizontal drainage pipe shall be in accordance with Table 704.1.*

Discussion and Commentary: Optimal drainage piping performance depends on the pipe size and slope, both of which influence the waste velocity within the drain pipe system. If the velocity is too low in a drain pipe that is excessively oversized, the solids tend to drop out of suspension, settling to the bottom of the pipe. This may eventually result in a drain stoppage. A greater velocity is required to move solids at rest than is required to keep moving solids in suspension.

TABLE 704.1
SLOPE OF HORIZONTAL DRAINAGE PIPE

SIZE (inches)	MINIMUM SLOPE (inch per foot)
$2^{1}/_{2}$ or less	$^{1}/_{4}$
3 to 6	$^{1}/_{8}$
8 or larger	$^{1}/_{16}$

For SI: 1 inch = 25.4 mm, 1 inch per foot = 0.083.3 mm/m.

The minimum desired velocity in a horizontal drain pipe is approximately 2 feet per second (0.61 m/s). This velocity is often referred to as the *scouring velocity*. The scouring velocity is intended to keep solids in suspension.

Code Text: *The size of the drainage piping shall not be reduced in size in the direction of the flow. A 4-inch by 3-inch (102 mm by 76 mm) water closet connection shall not be considered as a reduction in size.*

Discussion and Commentary: One of the fundamental requirements of a drainage system is that piping cannot be reduced in size in the direction of drainage flow. A size reduction would create an obstruction to flow, possibly resulting in a backup of flow, an interruption of service in the drainage system or stoppage in the pipe. A 4-inch by 3-inch (102 mm by 76 mm) water closet connection is not considered a reduction in pipe size.

Prohibited installation:
Improper fittings and reduction in size

DRAINAGE SYSTEM: Piping within a public or private premise that conveys sewage, rainwater or other liquid wastes to a point of disposal. A drainage system does not include the mains of a public sewer system or a private or public sewage treatment or disposal plant.

Code Text: *Horizontal branches shall connect to the bases of stacks at a point located not less than 10 times the diameter of the drainage stack downstream from the stack. Except as prohibited by Section 711.2, horizontal branches shall connect to horizontal stack offsets at a point located not less than 10 times the diameter of the drainage stack downstream from the upper stack.*

Discussion and Commentary: At the base of every stack, a phenomenon of flow may occur that is commonly called *hydraulic jump.* Hydraulic jump is the rising in the depth of flow above half full. A horizontal drain designed in accordance with the code is expected to have a normal flow depth of not greater than half full. The rise in flow may be great enough to close off the opening of the pipe, thus creating pressure fluctuations in the system that may affect trap seals. Hydraulic jump occurs within a distance of 10 times the diameter of the drainage stack downstream of the stack connection. To avoid flow interference, backup of flow and extreme pressure fluctuations, horizontal branch connections are prohibited in this area.

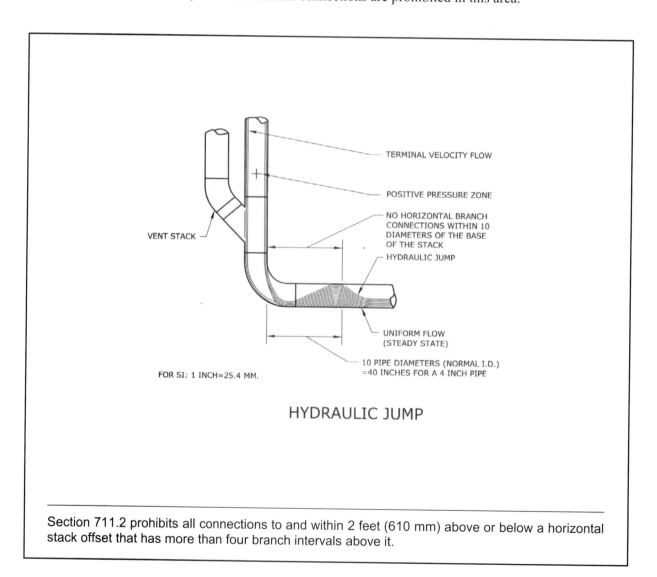

TERMINAL VELOCITY FLOW

POSITIVE PRESSURE ZONE

NO HORIZONTAL BRANCH CONNECTIONS WITHIN 10 DIAMETERS OF THE BASE OF THE STACK

HYDRAULIC JUMP

VENT STACK

UNIFORM FLOW (STEADY STATE)

10 PIPE DIAMETERS (NORMAL I.D.) =40 INCHES FOR A 4 INCH PIPE

FOR SI: 1 INCH=25.4 MM.

HYDRAULIC JUMP

Section 711.2 prohibits all connections to and within 2 feet (610 mm) above or below a horizontal stack offset that has more than four branch intervals above it.

Code Text: *Joints between polyethylene plastic pipe and fittings shall be underground and shall comply with Section 705.16.1 or 705.16.2.*

Discussion and Commentary: Polyethylene pipe joints are required to be covered for protection from damage. The heat fusion and mechanical joints are each covered in Sections 705.16.1 and 705.16.2.

ASTM

ASTM International
100 Barr Harbor Drive
West Conshohocken, PA 19428-2959

Section	Material	Application	Standard	Title	Applies to IPC Sections
705.16	polyethylene (PE) plastic pipe (SDR-PR)	building sewer pipe	ASTM F 714	Specification for Polyethylene (PE) Plastic Pipe (SDR-PR) based on Outside Diameter	Table 702.3

Polyethylene pipe is a plastic pipe that comes in various densities and is manufactured from petroleum hydrocarbons.

Code Text: *Joint surfaces shall be clean and free from moisture. All joint surfaces shall be cut, heated to melting temperature and joined using tools specifically designed for the operation. Joints shall be undisturbed until cool. Joints shall be made in accordance with ASTM D 2657 and the manufacturer's instructions.*

Discussion and Commentary: To obtain the best and most reliable joints, the surface of joint areas must be clean from debris and other matter and must also be dry. Heat fusion joints are not allowed to be created without using special tools intended for heat fusion.

ASTM

ASTM International
100 Barr Harbor Drive
West Conshohocken, PA 19428-2959

Section	Material	Application	Standard	Title	Applies to IPC Sections
705.16.1	polyethylene (PE) plastic pipe (SDR-PR)	heat-fusion joints for underground building sewer pipe	ASTM D 2657	Standard Practice for Heat Fusion-joining of Polyolefin Pipe and Fitting	605.19.2, 605.20.2, 705.16.1

Requirements, procedures and instructions from both ASTM D 2657 (*Standard Practice for Heat-Fusion-joining of Polyolefin Pipe and Fitting*) and the pipe manufacturer must be followed.

Topic: Mechanical Joints

Reference: IPC 705.16.2

Category: Sanitary Drainage

Subject: Polyethylene Plastic Pipe

Code Text: *Mechanical joints in drainage piping shall be made with an elastomeric seal conforming to ASTM C 1173, ASTM D 3212 or CSA B602. Mechanical joints shall be installed in accordance with the manufacturer's instructions.*

Discussion and Commentary: Elastomeric seals are required to create mechanical joints in the sanitary drainage system. These seals have been designed and are tested for reliable performance. Several standards are available for the standardization of elastomeric seals, some of which are: ASTM C 1173 (*Specification for Flexible Transition Couplings for Underground Piping Systems*) , ASTM D 3212 (*Specification for Joints for Drain and Sewer Plastic Pipes Using Flexible Elastomeric Seals*) and CSA-B 602 (*Mechanical Couplings for Drain, Waste and Vent Pipe and Sewer Pipe*).

ASTM

ASTM International
100 Barr Harbor Drive
West Conshohocken, PA 19428-2959

Section	Material	Application	Standard	Title	Applies to IPC Sections
705.16.2	Polyethylene (PE) plastic pipe (SDR-PR)	Mechanical and compression sleeve joints for underground building sewer pipe	ASTM C 1173	Specification for Flexible Transition Couplings for Underground Piping System	705.7.1, 705.14.1, 705.16
			ASTM D 3212	Specification for Joints for Drain and Sewer Plastic Pipes Using Flexible Elastomeric Seals	705.2.1, 705.7.1, 705.8.1, 705.14.1, 705.16.2

The installation of mechanical joints must always be in accordance with the manufacturer's instructions for proper performance and warranty purposes.

Code Text: *Heat-fusion joints for polyolefin pipe and tubing joints shall be installed with socket-type heat-fused polyolefin fittings or electrofusion polyolefin fittings. Joint surfaces shall be clean and free from moisture. The joint shall be undisturbed until cool. Joints shall be made in accordance with ASTM F 1412 or CSA B181.3.*

Discussion and Commentary: Of the various heat fusion procedures, only the socket-type and electrofusion fittings are allowed for polyolefin pipe and tubing in the sanitary drainage system. Such joints must comply with either ASTM F 1412 (*Specification for Polyolefin Pipe and Fittings for Corrosive Waste Drainage*) or CSA-B 181.3 (*Polyolefin Laboratory Drainage Systems*).

Polyolefin pipe and fittings for drainage systems are produced of schedule 40 and 80 materials.

Code Text: *Mechanical and compression sleeve joints shall be installed in accordance with the manufacturer's instructions.*

Discussion and Commentary: Because there are various types of mechanical joints that can be used in the sanitary drainage system, there is not any prescriptive methodology described for them in the code. As such, these joints must be installed according to the instructions of the manufacturer, which ensures performance and is needed for warranty purposes.

MECHANICAL JOINT: A connection between pipes, fittings, or pipes and fittings that is not screwed, caulked, threaded, soldered, solvent cemented, brazed or welded. A joint in which compression is applied along the centerline of the pieces being joined. In some applications, the joint is part of a coupling, fitting or adapter.

Code Text: *Fittings shall be installed to guide sewage and waste in the direction of flow. Change in direction shall be made by fittings installed in accordance with Table 706.3. Change in direction by combination fittings, side inlets or increasers shall be installed in accordance with Table 706.3 based on the pattern of flow created by the fitting. Double sanitary tee patterns shall not receive the discharge of back-to-back water closets and fixtures or appliances with pumping action discharge.*

Exception: Back-to-back water closet connections to double sanitary tees shall be permitted where the horizontal developed length between the outlet of the water closet and the connection to the double sanitary tee pattern is 18 inches (457 mm) or greater.

Discussion and Commentary: Drainage fittings and connections must provide a smooth transition of flow without creating obstructions or causing interference. The use of proper fittings helps to maintain the required flow velocities and reduces the possibility of stoppage in the drainage system. Combination fittings are commonly used in drainage systems and must be evaluated for their pattern of flow. Combination fittings include two or more fittings, such as combination wye and eighth bends or tee-wyes. A double sanitary tee (cross) cannot be used for connections to fixtures and appliances with pumping action, as such a fitting has a short pattern for change of direction.

If back-to-back fixtures discharge through several feet of pipe or multiple changes of direction, the flow in the fixture drains would decrease in velocity to the point where cross flow would not occur. The exception provides for back-to-back water closet connections to double sanitary tee patterns where the horizontal developed length between the outlet of the water closet and connection to the double sanitary tee pattern is 18 inches (457 mm) or greater.

TABLE 706.3
FITTINGS FOR CHANGE IN DIRECTION

TYPE OF FITTING PATTERN	CHANGE IN DIRECTION		
	Horizontal to vertical	Vertical to horizontal	Horizontal to horizontal
Sixteenth bend	X	X	X
Eighth bend	X	X	X
Sixth bend	X	X	X
Quarter bend	X	X[a]	X[a]
Short sweep	X	X[a,b]	X[a]
Long sweep	X	X	X
Sanitary tee	X[c]	—	—
Wye	X	X	X
Combination wye and eighth bend	X	X	X

For SI: 1 inch = 25.4 mm.
a. The fittings shall only be permitted for a 2-inch or smaller fixture drain.
b. Three inches or larger.
c. For a limitation on double sanitary tees, see Section 706.3.

Back-to-back water closets are fixtures that are installed directly opposite of each other and discharge through a short distance into a double pattern fitting.

Code Text: *Heel-inlet quarter bends shall be an acceptable means of connection, except where the quarter bend serves a water closet. A low-heel inlet shall not be used as a wet-vented connection. Side-inlet quarter bends shall be an acceptable means of connection for drainage, wet venting and stack venting arrangements.*

Discussion and Commentary: Low heel inlets on a quarter bend can become flooded and therefore should not be utilized as wet-vented connections to the drainage system. Because side inlet quarter bends normally do not become flooded on branches without water closets they can be utilized as a wet vented connection.

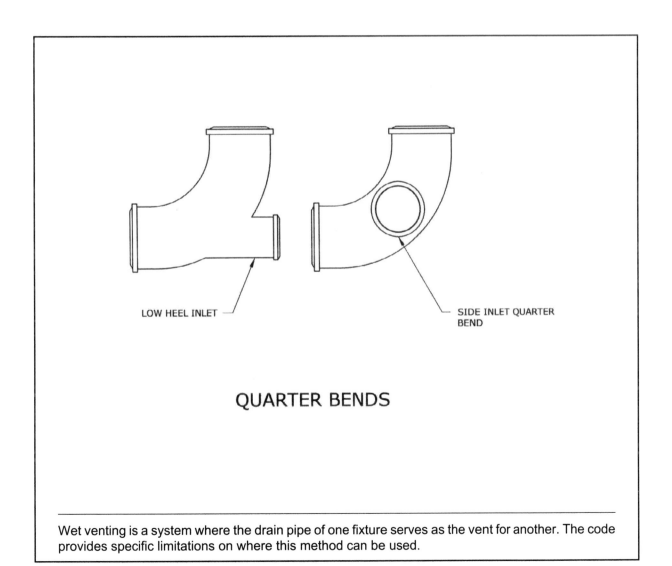

LOW HEEL INLET

SIDE INLET QUARTER BEND

QUARTER BENDS

Wet venting is a system where the drain pipe of one fixture serves as the vent for another. The code provides specific limitations on where this method can be used.

Quiz

Study Session 9
IPC Sections 701 – 707

1. When is a common building sewer connecting more than one building to the public sewer allowed?

   a. when the building tenants agree to the shared use of the sewer

   b. when the city sewer system does not extend to the property line

   c. when the common sewer is controlled by a property owners association

   d. when all buildings using the common sewer are located on the same property

   Reference _____

2. Which of the following waste products are prohibited from being discharged to the building drainage system?

   a. animal waste from a dog kennel

   b. liquid waste from a mortuary embalming room

   c. wastewater at temperatures exceeding 140°F

   d. backwash water from a swimming pool filtration system

   Reference _____

3. Exposed overhead soil or waste piping is prohibited from which of the following locations?

    a. medical clinic exam rooms

    b. restaurant supply warehouses

    c. soiled linen utility rooms in hospitals

    d. storage pantries in food service establishments

Reference _____

4. All of the following materials are approved for both above ground and below ground installations of the building drainage system *except* _____ .

    a. polyolefin   b. polyethylene

    c. polyvinyl chloride   d. stainless steel

Reference _____

5. Which of the following restrictions is placed on the installation of chemical waste piping?

    a. Only metallic piping is approved.

    b. The piping must be suitable for all concentrations of corrosive chemicals.

    c. The piping system must be completely separate from the sanitary system.

    d. The piping system must be a minimum of schedule 80 weight pipe.

Reference _____

6. Which of the following fittings is approved for the installation of polyethylene plastic pipe conforming to ASTM F 714?

    a. threaded fittings   b. caulked joint fittings

    c. solvent cemented fittings   d. mechanical joint fittings

Reference _____

7. Which of the following is required in the make-up of a solvent cement joint in 3-inch PVC DWV piping?

    a. the application of a purple primer conforming to ASTM F 656

    b. the application of solvent cement conforming to ASTM F 493

    c. the application of solvent cement that is yellow in color

    d. the application of solvent cement that is purple in color

Reference _____

8. Which of the following joints is approved for connecting copper pipe to galvanized pipe?

    a.   threaded copper adapter   b.   threaded brass adapter

    c.   threaded galvanized fitting   d.   soldered copper fitting

Reference _____

9. Which of the following joints is approved for connecting stainless steel drainage pipe to other material types?

    a.   welded joints   b.   threaded fittings

    c.   mechanical couplings   d.   red brass caulking ferrules

Reference _____

10. Which of the following pipe materials is not suitable for joining by a heat fusion method?

    a.   polyethylene   b.   polyolefin

    c.   polypropylene   d.   polyvinyl chloride

Reference _____

11. Which fixture is prohibited from connection to a fixture branch by double sanitary tee pattern fittings?

    a.   laundry sink   b.   kitchen sink

    c.   automatic clothes washer   d.   combination tub/shower

Reference _____

12. A 2-inch fixture drain connection, from the vertical to the horizontal direction of flow, is prohibited from connecting by use of a _____ .

    a.   quarter bend   b.   sanitary tee

    c.   short sweep fitting   d.   combination wye and eighth bend

Reference _____

13. Which fixture is prohibited from discharging through a quarter bend that has a heel inlet?

    a.   lavatory   b.   shower

    c.   urinal   d.   water closet

    Reference _____

14. The minimum required slope for a fixture drain from an automatic clothes washer is:

    a. $^1/_{16}$ inch per foot   b. $^1/_8$ inch per foot

    c. $^1/_4$ inch per foot   d. $^1/_2$ inch per foot

    Reference _____

15. Horizontal branches shall connect to horizontal stack offsets at a point located not less than _____ stack diameters downstream from the upper stack.

    a.   10   b.   20

    c.   30   d.   40

    Reference _____

16. A storm sewer must be separated from the building sewer by not less than _____ feet of compacted fill or undisturbed earth.

    a.   0   b.   2

    c.   4   d.   5

    Reference _____

17. A 4-inch by 3-inch water closet connection is _____ .

    a.   not considered a reduction in size

    b.   prohibited

    c.   an offset fitting

    d.   required

    Reference _____

18. Cleanout extensions are _____ .

    a. dead ends   b. limited to 12 inches in length

    c. prohibited   d. not considered dead ends

Reference _____

19. Solvent-cement joints for ABS plastic fittings are permitted _____ .

    a. above or below ground   b. only above ground

    c. only below ground   d. below ground if approved

Reference _____

20. Mechanical joints in PVC pipe are permitted _____ .

    a. above or below ground   b. only above ground

    c. only below ground   d. below ground if approved

Reference _____

21. A _____ joint is permitted to join plastic pipe and cast-iron hub pipe.

    a. TFE seal   b. caulked

    c. slip   d. fitting adapter

Reference _____

22. Caulking ferrules shall be _____ .

    a. red brass   b. bronze

    c. pvc   d. stainless steel

Reference _____

23. _____ shall be of the recessed drainage type.

    a. Saddle-type fittings

    b. Cement or concrete joints

    c. Threaded drainage pipe fittings

    d. Coextruded pipe fittings

Reference _____

24. A _____ is not permitted for a vertical to horizontal change of direction.

    a. sixth bend

    b. sanitary tee

    c. long sweep

    d. combination wye and eighth bend

Reference _____

25. For a 3-inch diameter branch drain, a _____ is permitted for a horizontal to horizontal change of direction.

    a. quarter bend

    b. sanitary tee

    c. short sweep

    d. combination wye and eighth bend

Reference _____

2006 IPC Sections 708 – 715
Sanitary Drainage II

OBJECTIVE: To develop an understanding of the overall code provisions that regulate the materials, design and installation of sanitary drainage. To develop an understanding of the specific provisions of the code that apply to cleanouts, sumps and ejectors.

REFERENCE: Sections 708 – 715, 2006 *International Plumbing Code*

KEY POINTS:
- What material is required for cleanout plugs?
- Cleanouts are required at what spacing for horizontal drains?
- What spacing is required for cleanouts on building sewers, and where is it measured?
- When is it permitted to omit the cleanout at the junction of the sewer and drain?
- Whenever the crawlspace is less than 24 inches in height, what provisions are required for the cleanout?
- Is it permitted to use a cleanout opening to install a new fixture?
- A 3-inch pipe would require what size cleanout?
- What do fixture unit values designate, and how are they used?
- What is the minimum trap size for an unlisted fixture?
- For continuous flow and semicontinuous flow, two fixture units would represent how many gallons per minute (gpm)?
- What determines the sum of drainage fixture units?
- What limitations are placed on the location of a horizontal branch in relation to a horizontal stack?
- When building drains cannot discharge to the sewer by gravity, what is required?

Code Text: *Cleanout plugs shall be brass or plastic, or other approved materials. Brass cleanout plugs shall be utilized with metallic drain, waste and vent piping only, and shall conform to ASTM A 74, ASME A112.3.1 or ASME A112.36.2M. Cleanouts with plate-style access covers shall be fitted with corrosion-resisting fasteners. Plastic cleanout plugs shall conform to the requirements of Section 702.4. Plugs shall have raised square or countersunk square heads. Countersunk heads shall be installed where raised heads are a trip hazard. Cleanout plugs with borosilicate glass systems shall be of borosilicate glass.*

Discussion and Commentary: Metallic cleanout plugs must be brass to provide for easy removal. If the cleanout plug is corroded in place, the brass plug is a soft enough material to be chiseled out. Brass cleanout plugs are limited to metallic fittings because the metal plug threads may damage the softer plastic threads of a fitting. Plastic plugs are intended for use with plastic fittings; however, this section does not prohibit the use of plastic plugs with metallic fittings. Like brass plugs, plastic plugs are less likely to seize in metallic fittings. The cleanout plug must have a square turning surface to allow for ease of removal while minimizing the possibility of stripping the surface during removal.

The intent of the code is to require threaded cleanout plugs to be constructed of either brass or plastic. Specialized borosilicate glass drainage piping requires the use of a compatible borosilicate glass cleanout plug to maintain corrosion resistance.

Code Text: *All horizontal drains shall be provided with cleanouts located not more than 100 feet (30 480 mm) apart.*

Discussion and Commentary: Cleanouts must be spaced a reasonable distance from each other to facilitate cleaning any portion of the drainage system. This distance is based on the use of modern-day cleaning equipment and attempts to minimize the inconvenience and health hazard associated with creating a mess inside the building as a result of removing a blockage.

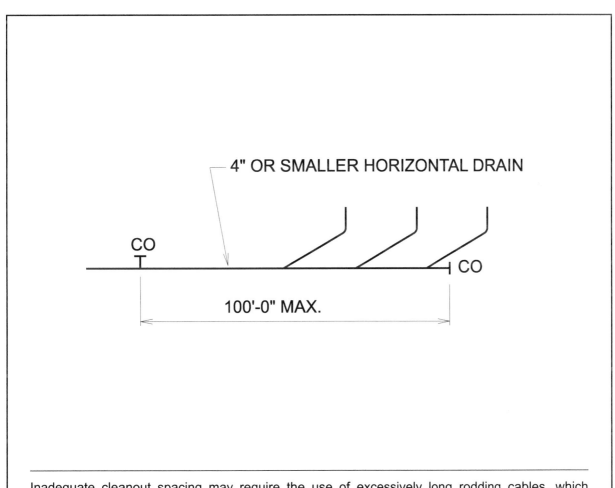

Inadequate cleanout spacing may require the use of excessively long rodding cables, which complicates the task and increases the likelihood that the rodding cable will jam or break inside the pipe. Every horizontal drain must have at least one cleanout, regardless of the length.

Code Text: *Building sewers shall be provided with cleanouts located not more than 100 feet (30 480 mm) apart measured from the upstream entrance of the cleanout. For building sewers 8 inches (203 mm) and larger, manholes shall be provided and located not more than 200 feet (60 960 mm) from the junction of the building drain and building sewer, at each change in direction and at intervals of not more than 400 feet (122 m) apart. Manholes and manhole covers shall be of an approved type.*

Discussion and Commentary: A distance of up to 100 feet (30 480 mm) is permitted between cleanouts on building sewers because drain-cleaning equipment may be used outdoors where this distance can be easily accommodated. The use of outdoor cleanouts does not involve the same concerns as indoor cleanouts relative to health hazards and protection of property. Outdoor cleanouts for sewers must be brought up to grade for access, and the length of piping between the actual cleanout access opening and the sewer must be included in the overall developed length of piping between cleanouts. When a building sewer is 8 inches (203 mm) or larger, manholes are required as cleanouts, and a distance of 200 feet (60 960 mm) is required from the junction of the building drain and the building sewer.

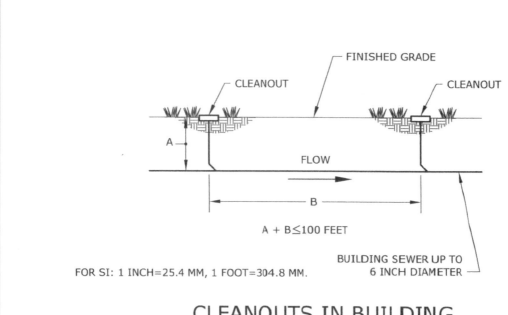

FINISHED GRADE

CLEANOUT

CLEANOUT

A

FLOW

B

A + B ≤ 100 FEET

FOR SI: 1 INCH=25.4 MM, 1 FOOT=304.8 MM.

BUILDING SEWER UP TO 6 INCH DIAMETER

CLEANOUTS IN BUILDING SEWER-LIMITATIONS

The manhole cover may be located above the base flood level elevation or designed to resist 1) hydrostatic forces when submerged in water and 2) dynamic forces that are due to wave action or high-velocity water.

Topic: Changes in Direction

Category: Sanitary Drainage

Reference: IPC 708.3.3

Subject: Cleanouts

Code Text: *Cleanouts shall be installed at each change of direction greater than 45 degrees (0.79 rad) in the building sewer, building drain and horizontal waste or soil lines. Where more than one change of direction occurs in a run of piping, only one cleanout shall be required for each 40 feet (12 192 mm) of developed length of the drainage piping.*

Discussion and Commentary: The requirement for a cleanout at a change in direction greater than 45 degrees (0.79 rad) is for an individual fitting, not a combination of fittings. If a 90-degree (1.6 rad) change in direction is accomplished with a single fitting, such as a sweep or combination tee-wye, a cleanout is required. If the same change in direction is accomplished with two one-eighth bends, a cleanout is not required.

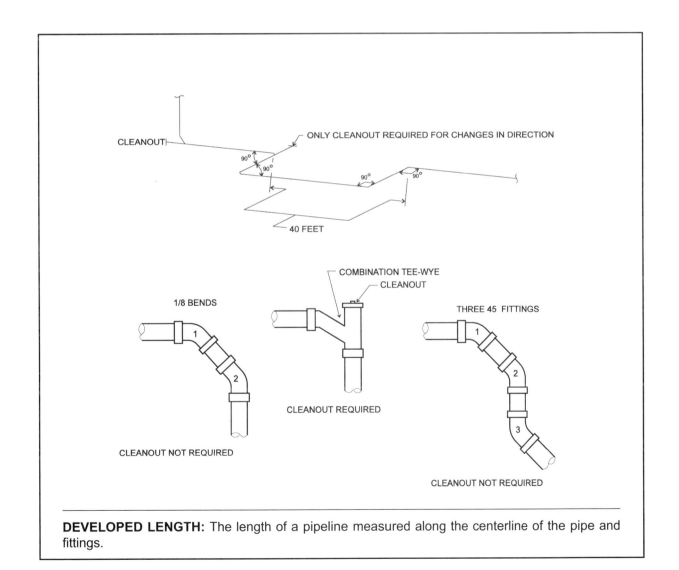

DEVELOPED LENGTH: The length of a pipeline measured along the centerline of the pipe and fittings.

Code Text: *A cleanout shall be provided at the base of each waste or soil stack.*

Discussion and Commentary: The characteristics of the drainage flow in horizontal pipe increase the probability of a stoppage. Additionally, it is possible for solids to collect at the change of direction from vertical to horizontal; therefore, a cleanout is required at the base of every drainage stack to provide access to the horizontal piping that serves the stack.

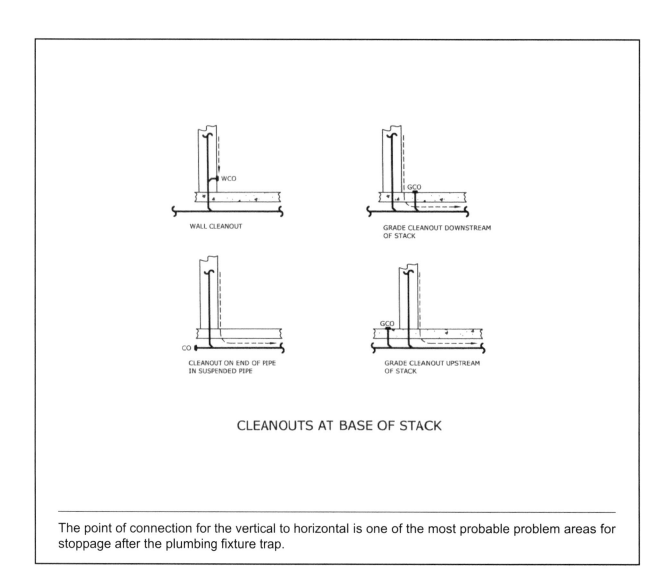

CLEANOUTS AT BASE OF STACK

The point of connection for the vertical to horizontal is one of the most probable problem areas for stoppage after the plumbing fixture trap.

Code Text: *Manholes serving a building drain shall have secured gas-tight covers and shall be located in accordance with Section 708.3.2.*

Discussion and Commentary: Where a manhole is provided to serve as a cleanout for a building drain, such manhole is required to have a secured, gas-tight cover. This requirement is to prevent the escape of sewer gas and to reduce the possibility of unauthorized access to the manhole. See Section 708.3.2 for additional requirements and concerns related to the installation of manholes.

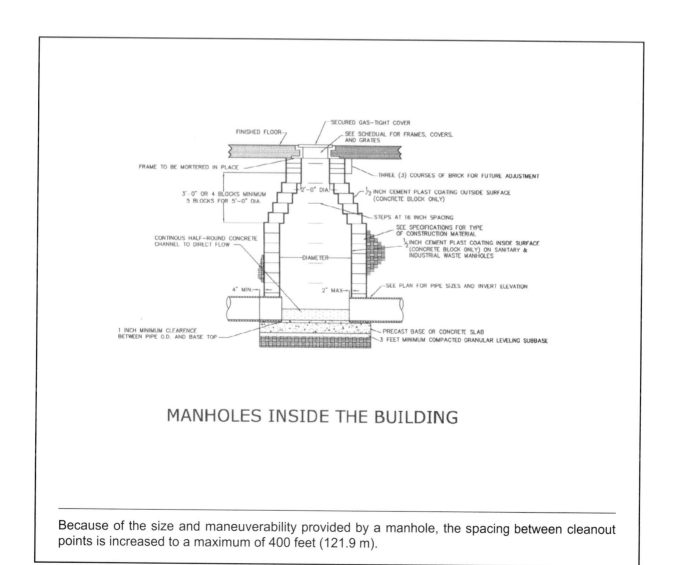

MANHOLES INSIDE THE BUILDING

Because of the size and maneuverability provided by a manhole, the spacing between cleanout points is increased to a maximum of 400 feet (121.9 m).

Code Text: *Cleanouts on concealed piping or piping under a floor slab or in a crawl space of less than 24 inches (610 mm) in height or a plenum shall be extended through and terminate flush with the finished wall, floor or ground surface or shall be extended to the outside of the building. Cleanout plugs shall not be covered with cement, plaster or any other permanent finish material. Where it is necessary to conceal a cleanout or to terminate a cleanout in an area subject to vehicular traffic, the covering plate, access door or cleanout shall be of an approved type designed and installed for this purpose.*

Discussion and Commentary: Because a cleanout is designed to provide access into the drainage system, the cleanout itself must be accessible, regardless of whether the piping it serves is concealed or in a location not readily accessed. The cleanout is required to extend up to and flush with the finished floor level or outside grade or to a location that is flush with a finished wall. Cleanouts in drains located above very high ceilings may require the cleanout to be turned up to the floor slab above or into a wall on the floor above. Cleanouts located on walking surfaces must be countersunk to minimize the tripping hazard and to protect the cleanout from damage.

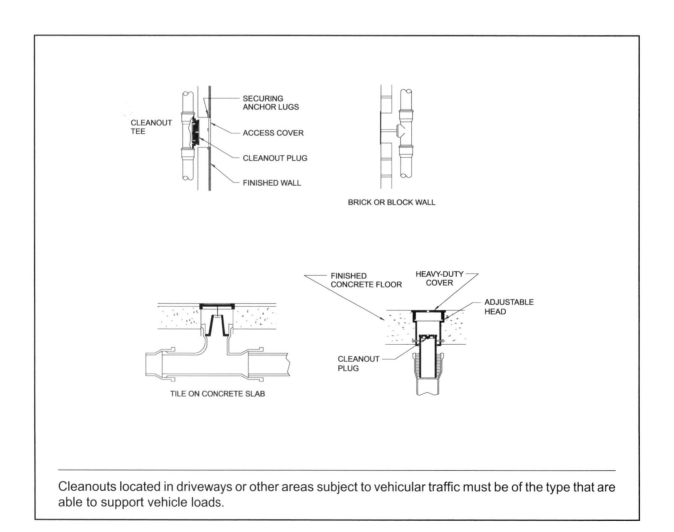

Cleanouts located in driveways or other areas subject to vehicular traffic must be of the type that are able to support vehicle loads.

Code Text: *Cleanout openings shall not be utilized for the installation of new fixtures, except where approved and where another cleanout of equal access and capacity is provided.*

Discussion and Commentary: In existing structures, cleanouts provide a convenient opening for the connection of new piping in remodeling, addition and alteration work. The cleanout fitting is commonly removed to allow a new connection; however, a substitute cleanout must be provided to serve in the same capacity as the cleanout that was eliminated. Many threaded cleanout openings have only a few threads and, therefore, are not intended to receive threaded pipe or male adapters.

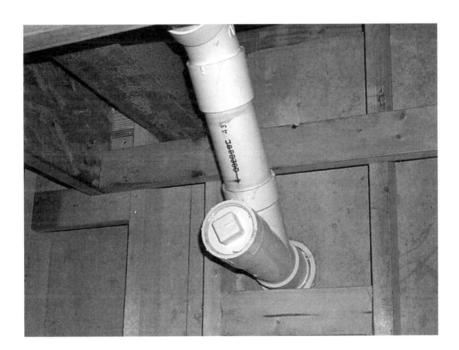

The practice of converting a floor cleanout to a floor drain is prohibited because floor drains require a trap seal to prevent sewer gases entering the building interior.

Code Text: *Cleanouts on 6-inch (153 mm) and smaller pipes shall be provided with a clearance of not less than 18 inches (457 mm) for rodding. Cleanouts on 8-inch (203 mm) and larger pipes shall be provided with a clearance of not less than 36 inches (914 mm) for rodding.*

Discussion and Commentary: This section provides minimum access clearances for cleanouts. The code is very specific on the requirements related to providing drainage piping cleanouts. The usability of these cleanouts is not to be compromised by their placement in the building. The clearances given herein are viewed as the minimum conditions that will allow the cleanout to serve its purpose. As such, the designer must attempt to provide additional clearance whenever possible.

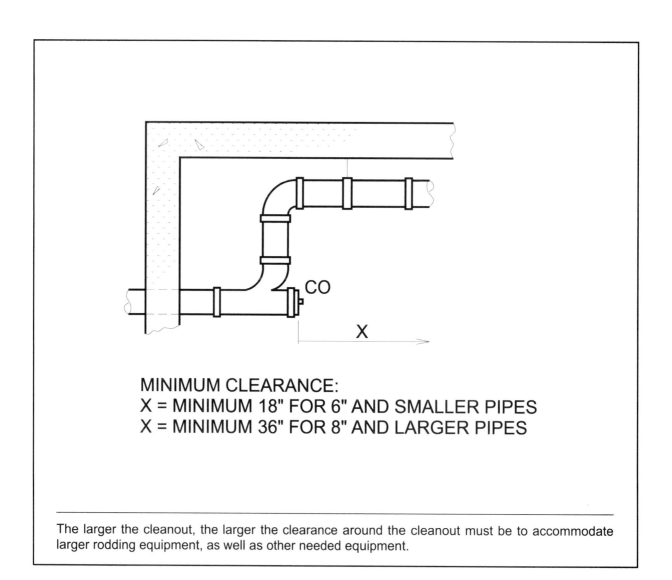

MINIMUM CLEARANCE:
X = MINIMUM 18" FOR 6" AND SMALLER PIPES
X = MINIMUM 36" FOR 8" AND LARGER PIPES

The larger the cleanout, the larger the clearance around the cleanout must be to accommodate larger rodding equipment, as well as other needed equipment.

Code Text: *Fixtures not listed in Table 709.1 shall have a drainage fixture unit load based on the outlet size of the fixture in accordance with Table 709.2. The minimum trap size for unlisted fixtures shall be the size of the drainage outlet but not less than 1.25 inches (32 mm).*

Discussion and Commentary: When a specific plumbing fixture is not listed in Table 709.1, the dfu value is based on the outlet size of the fixture. A $1^1/_4$-inch (32 mm) drain is the minimum acceptable size to permit proper open channel flow in the pipe for sanitary drainage. The minimum trap size is based on this pipe size.

TABLE 709.2
DRAINAGE FIXTURE UNITS FOR FIXTURE DRAINS OR TRAPS

FIXTURE DRAIN OR TRAP SIZE (inches)	DRAINAGE FIXTURE UNIT VALUE
$1^1/_4$	1
$1^1/_2$	2
2	3
$2^1/_2$	4
3	5
4	6

For SI: 1 inch = 25.4 mm.

Floor sinks (receptors) are not listed in Table 709.1; therefore, Table 709.2 would be used to determine their drainage fixture unit value.

Code Text: *Drainage fixture unit values for continuous and semicontinuous flow into a drainage system shall be computed on the basis that 1 gpm (0.06 L/s) of flow is equivalent to two fixture units.*

Discussion and Commentary: Equipment that discharges either continuously or semicontinuously, such as pumps, ejectors, air-conditioning equipment, commercial laundries and commercial dishwashers, are assigned two fixture units for each gpm of discharge rate. For example, a continuously operating pump with a discharge rate of 20 gpm (78 L/min) would have a dfu value of 40. Because a dfu is based on probability, only whole numbers are utilized to express fixture unit values.

The relationship of dfu to gpm is not a constant ratio that allows direct conversion of units. Thus, it cannot be determined that a dfu value of 40 would yield a continuous flow of 20 gpm (78 L/min).

Topic: Sump Pit

Reference: IPC 712.3.2

Category: Sanitary Drainage

Subject: Sumps and Ejectors

Code Text: *The sump pit shall be not less than 18 inches (457 mm) in diameter and 24 inches (610 mm) deep, unless otherwise approved. The pit shall be accessible and located such that all drainage flows into the pit by gravity. The sump pit shall be constructed of tile, concrete, steel, plastic or other approved materials. The pit bottom shall be solid and provide permanent support for the pump. The sump pit shall be fitted with a gas-tight removable cover adequate to support anticipated loads in the area of use. The sump pit shall be vented in accordance with Chapter 9.*

Discussion and Commentary: The minimum dimensions of 18 inches (457 mm) in diameter and 24 inches (610 mm) in depth are required, unless the designer or manufacturer can present calculations and other supporting data that will allow the code official to determine that other dimensions work for the situation being addressed. The sump pit is required to be constructed of a durable material, such as tile, concrete, steel, plastic or other approved materials. The bottom of the sump pit must be solid and structurally capable of supporting the sump pump. The sump pit is required to have a gas-tight removable cover to prevent the escape of sewer gas.

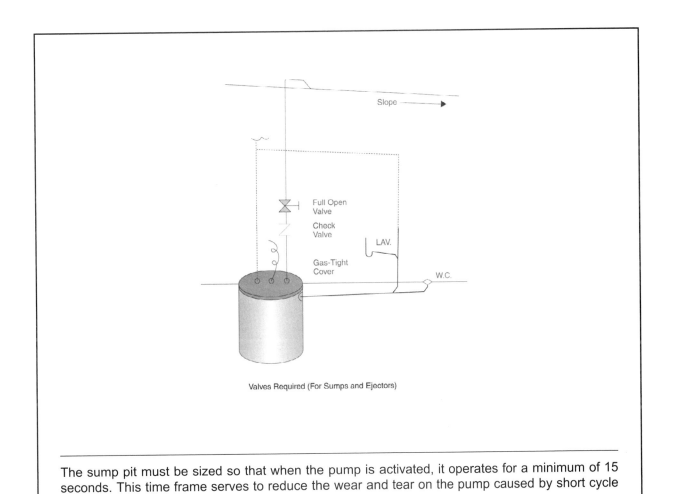

Valves Required (For Sumps and Ejectors)

The sump pit must be sized so that when the pump is activated, it operates for a minimum of 15 seconds. This time frame serves to reduce the wear and tear on the pump caused by short cycle operation.

Code Text: *Pumps connected to the drainage system shall connect to the building sewer or shall connect to a wye fitting in the building drain a minimum of 10 feet (3048 mm) from the base of any soil stack, waste stack or fixture drain. Where the discharge line connects into horizontal drainage piping, the connector shall be made through a wye fitting into the top of the drainage piping.*

Discussion and Commentary: The connection of the discharge piping must occur either directly to the building sewer or to a wye fitting in the building drain. If connected to the building drain, the wye fitting must be a minimum of 10 feet (3048 mm) away from other pipe connections or fixtures. This is to reduce the likelihood that discharge from the pump/ejector will interfere with the gravity flow in the drainage system. The 10-foot (3048 mm) distance will allow the pumped waste flow to settle in the invert of the building drain or sewer without creating backups and pressure surges.

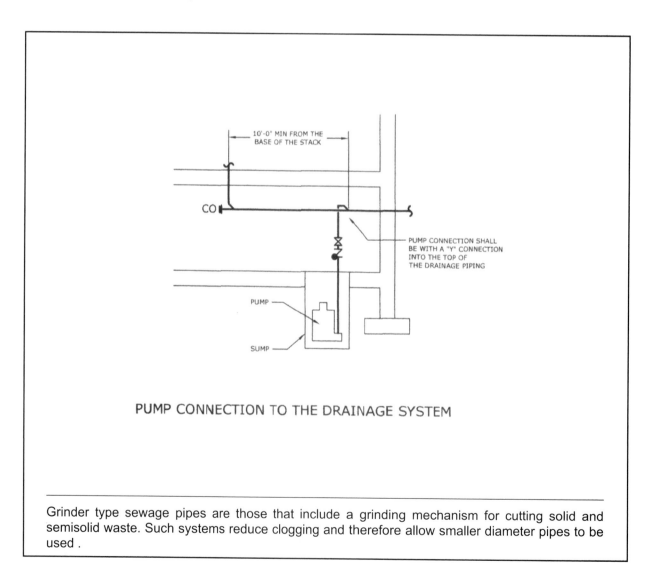

PUMP CONNECTION TO THE DRAINAGE SYSTEM

Grinder type sewage pipes are those that include a grinding mechanism for cutting solid and semisolid waste. Such systems reduce clogging and therefore allow smaller diameter pipes to be used .

Code Text: *Ready access shall be provided to vacuum system station receptacles. Such receptacles shall be built into cabinets or recesses and shall be visible.*

Discussion and Commentary: Medical vacuum system station inlets are typically installed in recesses in locations where medical professionals have ready access to them. The person locating the station must take into consideration the movements of personnel during an operation or emergency procedure. These stations are to be visible to allow for ready supervision of the equipment to ensure it is functioning as required.

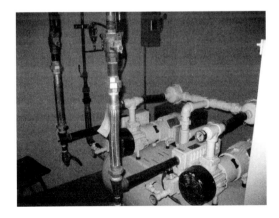

Medical vacuum outlets typically have a slide plate adjacent to the outlet so as to allow the mounting of a medical vacuum bottle collector, which serves as an interceptor for solids and liquids. This ensures that solids and liquids do not get suctioned into the piping system.

Code Text: *Where the flood level rims of plumbing fixtures are below the elevation of the manhole cover of the next upstream manhole in the public sewer, such fixtures shall be protected by a backwater valve installed in the building drain, branch of the building drain or horizontal branch serving such fixtures. Plumbing fixtures having flood level rims above the elevation of the manhole cover of the next upstream manhole in the public sewer shall not discharge through a backwater valve.*

Discussion and Commentary: A backwater valve is required in areas where the public sewer may back up into the building through the sanitary drainage system. When plumbing fixtures are located above the next upstream manhole cover from the building sewer connection to the public sewer, the sewer will back up through the street manhole before entering the building. Public sewers may become blocked or overloaded, which will result in sewage backing up into manholes and any laterals (taps) connected to the sewer system.

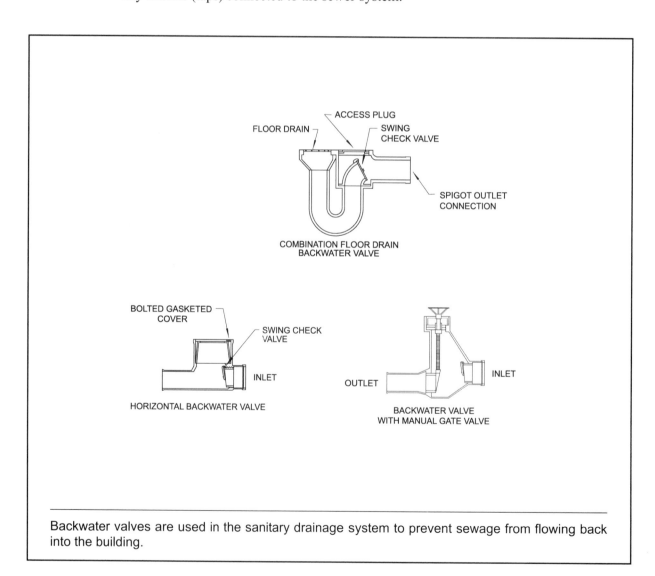

Backwater valves are used in the sanitary drainage system to prevent sewage from flowing back into the building.

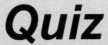

Quiz

Study Session 10
IPC Sections 708 – 715

1. Which of the following joints is a prohibited connection?

    a. saddle-type piercing valve for an ice maker line

    b. approved adapter fitting between different types of plastic

    c. mechanical coupling between stainless steel and other piping

    d. a caulked joint between a PVC closet bend and a cast-iron hub fitting

    Reference _____

2. A horizontal building drain is 80 feet in developed length with three 90 degree horizontal changes in direction located 12 feet, 40 feet and 60 feet from the junction with the building sewer. Cleanouts are required at the junction with the building sewer and at _____ feet from the junction with the building sewer.

    a. 12, 40 and 60   b. 12 and 60

    c. 40 and 60   d. 60

    Reference _____

3. For a two-inch horizontal branch drain, a _____ one pipe size smaller than the drain it serves is permitted as a cleanout.

    a. vent terminal

    b. a vertical connection to a sanitary tee

    c. P trap connection with slip joints

    d. a connection to a wye fitting

Reference _____

4. The drainage fixture unit load from a clinical sink with a 3-inch fixture outlet is _____ .

    a. 1   b. 2

    c. 3   d. 5

Reference _____

5. A sump pump with a discharge capacity of 10 GPM that discharges to the drainage system through a 2-inch indirect waste receptor will have a computed drainage fixture unit value of _____ .

    a. 3   b. 10

    c. 20   d. 42

Reference _____

6. A private bathroom with a 1.6 gpf water closet, lavatory, shower, bidet and floor drain has a total drainage fixture unit value of _____ .

    a. 5   b. 7

    c. 9   d. 10

Reference _____

7. The minimum size of any building drain serving a water closet is _____ inches.

    a. 3   b. 4

    c. 2   d. $2^{1}/_{2}$

Reference _____

8. A vertical distance of _____ feet or more between the connections of horizontal branches to a drainage stack is a branch interval.

    a. 7   b. 8

    c. 9   d. 10

    Reference _____

9. In a building with a basement subdrain sump, _____ is prohibited from discharging to the sump.

    a. subsoil drainage

    b. drainage from soil pipe

    c. waste from laundry fixtures

    d. drainage above the gravity sewer

    Reference _____

10. A _____ is not required for a sewage sump and ejector.

    a. backup pump or pump alarm system

    b. vent

    c. check valve

    d. full open valve

    Reference _____

11. The pump shall be set to prevent the effluent in the sump from rising to within _____ inches of the invert of the gravity drain inlet.

    a. $1^1/_2$ b. 2

    c. 3   d. 4

    Reference _____

12. The minimum dimensions of a sump pit are _____ inches in diameter and _____ inches deep.

    a. 16, 20   b. 18, 20

    c. 18, 24   d. 20, 24

Reference _____

13. The discharge pipe of a sewage ejector connecting to horizontal drainage piping shall be connected through a _____ .

    a. wye fitting into the side of the drainage piping

    b. sanitary tee into the side of the drainage piping

    c. sanitary tee into the top of the drainage piping

    d. wye fitting into the top of the drainage piping

Reference _____

14. Sewage grinder pumps that receive the discharge of water closets shall have a minimum discharge opening of _____ inch(es).

    a. .75   b. 1.5

    c. 1.25   d. 2

Reference _____

15. What is the minimum required capacity of a 2-inch sewage pump?

    a. 15 gpm   b. 21 gpm

    c. 30 gpm   d. 46 gpm

Reference _____

16. A _____ shall be vented to open air with a minimum 2-inch diameter vent.

    a. sterilizer   b. clinical sink

    c. bedpan washer   d. garbage can washer

Reference _____

17. Which of the following health care fixtures is required to discharge through an air gap into an indirect waste receptor?

    a.   clinical sink   b.   bedpan washer

    c.   bedpan steamer   d.   vacuum (fluid suction) system

Reference _____

18. What is the minimum required slope of a 2-inch horizontal drain when part of an approved computerized drainage system design?

    a. $^1/_{16}$ inch per foot   b. $^1/_8$ inch per foot

    c. $^1/_4$ inch per foot   d. $^1/_2$ inch per foot

Reference _____

19. A fully open backwater valve shall have a flow capacity not less than _____ percent of the pipe size served.

    a.   125   b.   100

    c.   85   d.   75

Reference _____

20. A backwater valve is required to protect plumbing fixtures with flood level rims below the elevation of the _____ .

    a.   manhole cover of the next upstream manhole

    b.   building sewer

    c.   building drain

    d.   public sewer

Reference _____

21. The air flow velocity of a central vacuum (fluid suction) system shall be less than _____ feet per minute.

    a.   4,000   b.   5,000

    c.   6,000   d.   3,000

Reference _____

22. All horizontal drains shall be provided with cleanouts located not more than _____ feet apart.

        a. 40   b. 60

        c. 80   d. 100

Reference _____

23. Where more than one change of direction occurs in a run of piping, only one cleanout shall be required for each _____ feet of developed length.

        a. 40   b. 60

        c. 80   d. 100

Reference _____

24. A cleanout shall be provided at the base of each _____ .

        a. 90 degree angle   b. vent stack

        c. waste stack   d. drain stack

Reference _____

25. A minimum clearance of _____ inches is required in front of a cleanout for a 2-inch drainage pipe.

        a. 12   b. 18

        c. 24   d. 30

Reference _____

2006 IPC Chapter 8
Indirect/Special Waste

OBJECTIVE: To develop an understanding of the code provisions for indirect connections to the sanitary drainage system. To develop an understanding of how the code provisions for indirect connections prevent sewage from backing up into the indirect waste pipe. To develop an understanding of the code provisions regulating special waste that contains hazardous chemicals.

REFERENCE: Chapter 8, 2006 *International Plumbing Code*

KEY POINTS:
- What type of appliance or equipment is required to discharge through an indirect waste pipe?
- Equipment associated with food is required to discharge through an indirect waste pipe by what means?
- What is required for the waste line serving a floor drain that is located in an area subject to freezing?
- Devices and equipment discharging potable water to the building drainage system shall be through what type of waste pipe and through what separation?
- What type of protection is used for the discharge of wastewater from swimming pools?
- Devices and equipment discharging nonpotable water to the building drainage system shall be through an indirect waste pipe and through what type of protection?
- In lieu of using a deck-mounted air gap, what can be done to the waste line?
- What two methods are used for the discharge from a commercial dishwashing machine?
- Indirect waste piping is required to be trapped whenever the pipe exceeds what distance in developed length or what distance in overall length?
- What is the minimum distance between the indirect waste pipe and the flood level rim of the waste receptor?
- Where is an air break required?
- What limitations are placed on the location of waste receptors, and what type of access is required?
- What determines the size of a waste receptor?
- What height is required for the hub or pipe of an indirect waste receptor above the floor?

KEY POINTS: • Are standpipes required to be trapped?
(Cont'd) • Steam pipes and pipes with water over 140°F are required to discharge into what type of connection or device?

• How are materials that could destroy or injure a system or create toxic fumes or interfere with the sewage treatment process to be treated?

Code Text: *All devices, appurtenances, appliances and apparatus intended to serve some special function, such as sterilization, distillation, processing, cooling, or storage of ice or foods, and that discharge to the drainage system, shall be provided with protection against backflow, flooding, fouling, contamination and stoppage of the drain.*

Discussion and Commentary: This section requires appliances and specialized equipment to be protected against contamination resulting from backflow. Distillation devices may contain bacteria, and cooling systems may have chemicals added to control corrosion, algae growth and biocides to limit the growth of organic pathogens in water systems. It is important to protect the potable water supply from these contaminations.

PRESSURE INSTRUMENT WASHER STERILIZER: A pressure vessel fixture designed to both wash and sterilize instruments during the operating cycle of the fixture.

Code Text: *Food-handling equipment and clear-water waste shall discharge through an indirect waste pipe as specified in Sections 802.1.1 through 802.1.7. All health-care related fixtures, devices and equipment shall discharge to the drainage system through an indirect waste pipe by means of an air gap in accordance with this chapter and Section 713.3. Fixtures not required by this section to be indirectly connected shall be directly connected to the plumbing system in accordance with Chapter 7.*

Discussion and Commentary: The requirements for indirect wastes are described in Sections 802.1.1 through 802.1.7. Health-care plumbing fixtures, appliances and equipment that are required to discharge through an indirect waste are addressed in Section 713.3. Fixtures not required to be indirectly connected by Chapter 8 are required to be directly connected to the plumbing system in accordance with Section 301.3.

INDIRECT WASTE PIPE: A waste pipe that does not connect directly with the drainage system but that discharges into the drainage system through an air break or air gap into a trap, fixture, receptor or interceptor.

Code Text: *Equipment and fixtures utilized for the storage, preparation and handling of food shall discharge through an indirect waste pipe by means of an air gap.*

Discussion and Commentary: All food must be protected from possible contamination caused by the sanitary drainage system. The requirement for indirect waste connections extends to all storage, cooking and preparation equipment, including vegetable sinks, food washing sinks, refrigerated cases and cabinets, ice boxes, ice-making machines, steam kettles, steam tables, potato peelers, egg boilers, coffee urns and brewers, drink dispensers and similar types of equipment and fixtures. Food stored in walk-in coolers and freezers is specifically addressed in Section 802.1.2.

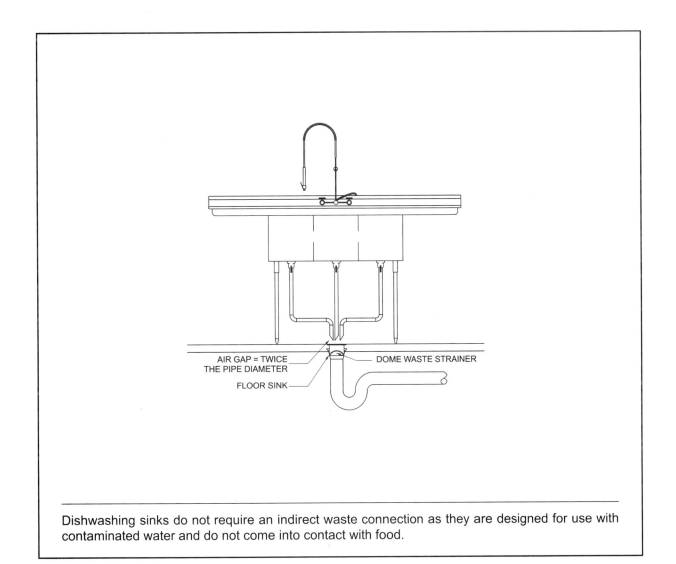

Dishwashing sinks do not require an indirect waste connection as they are designed for use with contaminated water and do not come into contact with food.

Code Text: *Where devices and equipment, such as sterilizers and relief valves, discharge potable water to the building drainage system, the discharge shall be through an indirect waste pipe by means of an air gap.*

Discussion and Commentary: An indirect waste connection by means of an air gap is required for potable clear water waste. An air gap prevents possible cross connection between potable water and the sanitary drainage system. Because open fixtures such as sinks cannot produce the negative pressures necessary for siphonage, backflow can only occur if waste is forced to rise in the fixture drain. Drains from closed vessels, such as water heater relief valve discharges, can cause backflow by siphonage, which an air break cannot prevent.

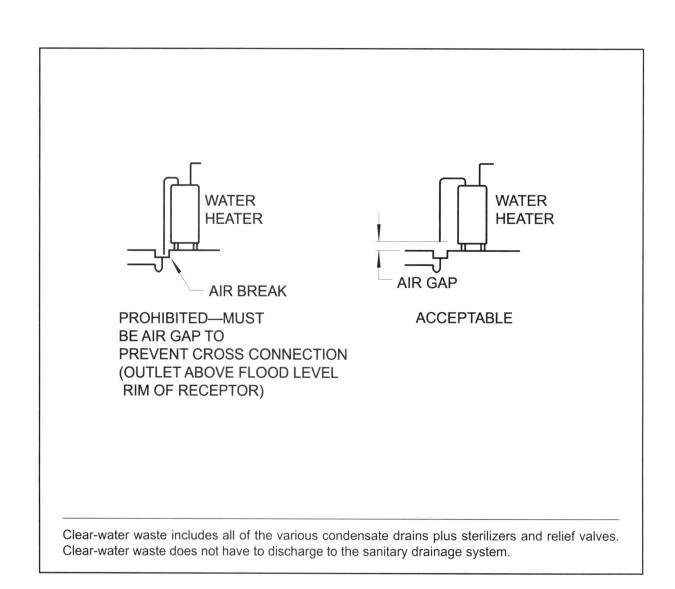

Clear-water waste includes all of the various condensate drains plus sterilizers and relief valves. Clear-water waste does not have to discharge to the sanitary drainage system.

Topic: Nonpotable Clear-water Waste

Category: Indirect/Special Waste

Reference: IPC 802.1.5

Subject: Indirect Wastes

Code Text: *Where devices and equipment such as process tanks, filters, drips and boilers discharge nonpotable water to the building drainage system, the discharge shall be through an indirect waste pipe by means of an air break or an air gap.*

Discussion and Commentary: Where there is no possibility of contaminating the potable water, the indirect waste may be by means of an air gap or air break. An air break is often preferred to reduce any splashing that may occur. This section does not require the devices and equipment listed to discharge to the drainage system; it only indicates the method of discharge if they do connect.

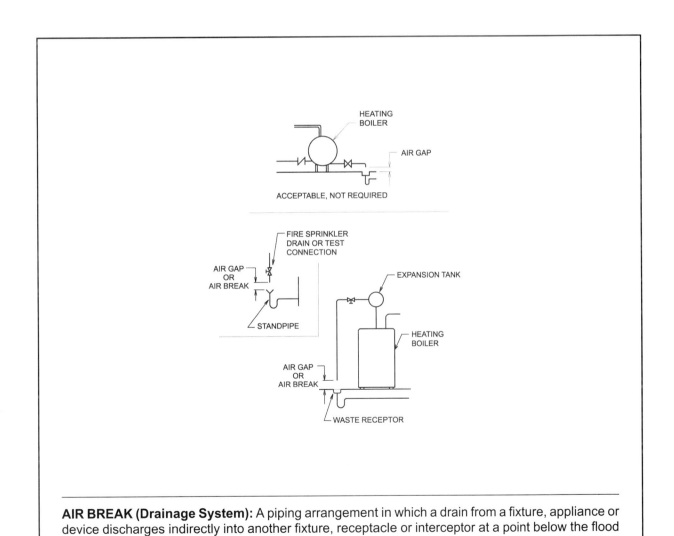

AIR BREAK (Drainage System): A piping arrangement in which a drain from a fixture, appliance or device discharges indirectly into another fixture, receptacle or interceptor at a point below the flood level rim and above the trap seal.

Code Text: *Domestic dishwashing machines shall discharge indirectly through an air gap or air break into a standpipe or waste receptor in accordance with Section 802.2, or discharge into a wye-branch fitting on the tailpiece of the kitchen sink or the dishwasher connection of a food waste grinder. The waste line of a domestic dishwashing machine discharging into a kitchen sink tailpiece or food waste grinder shall connect to a deck-mounted air gap or the waste line shall rise and be securely fastened to the underside of the sink rim or counter.*

Discussion and Commentary: Dishwashing machines must discharge indirectly to the drainage system. An indirect connection by air gap or air break is required. The waste line is required to be looped as high as possible and securely fastened to the underside of the sink rim or counter top. This is intended to minimize the potential for waste backflow into the dishwasher cabinet.

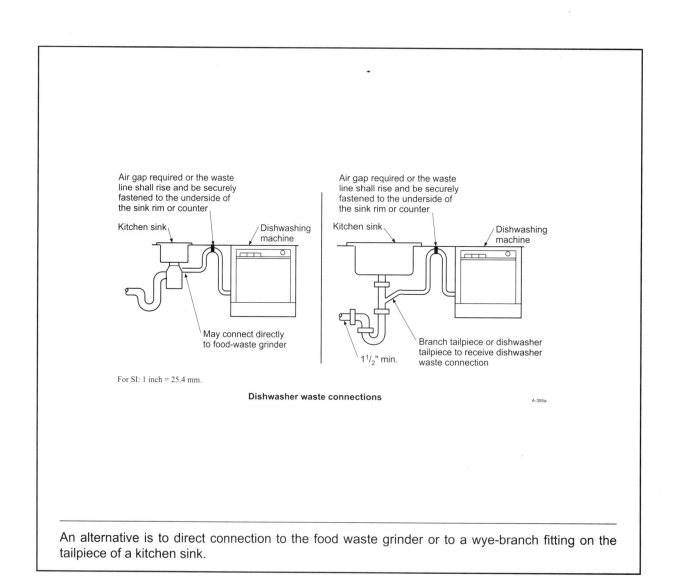

For SI: 1 inch = 25.4 mm.

Dishwasher waste connections

A-395a

An alternative is to direct connection to the food waste grinder or to a wye-branch fitting on the tailpiece of a kitchen sink.

Code Text: *All indirect waste piping shall discharge through an air gap or air break into a waste receptor or standpipe. Waste receptors and standpipes shall be trapped and vented and shall connect to the building drainage system. All indirect waste piping that exceeds 2 feet (610 mm) in developed length measured horizontally, or 4 feet (1219 mm) in total developed length, shall be trapped.*

Discussion and Commentary: To minimize bacterial growth and odor in indirect waste piping, a trap is required when the piping exceeds 2 feet (610 mm) in developed length measured horizontally or 4 feet (1219 mm) in total developed length. The trap requirement applies only to fixtures or devices that are open ended. It does not apply to an indirect waste pipe making a direct connection to a device such as a water heater relief valve.

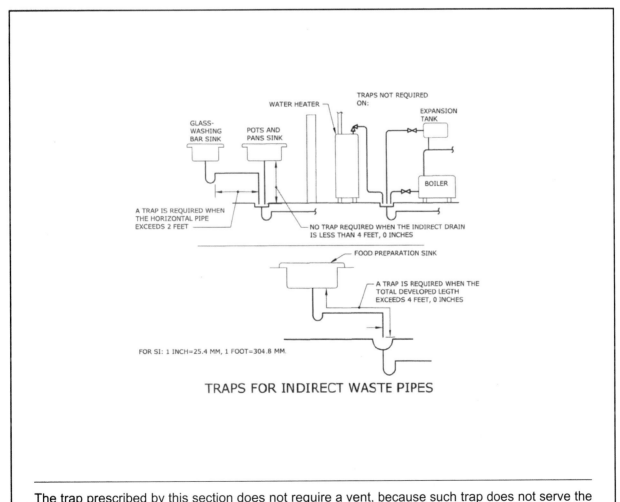

TRAPS FOR INDIRECT WASTE PIPES

The trap prescribed by this section does not require a vent, because such trap does not serve the same purpose as other traps in the drainage system.

Code Text: *The air gap between the indirect waste pipe and the flood level rim of the waste receptor shall be a minimum of twice the effective opening of the indirect waste pipe.*

Discussion and Commentary: An air gap is a separation of the drainage pipe that terminates above the flood level rim of the waste receptor. To be completely effective in accomplishing its intended purpose, the physical separation from the waste pipe to the flood level rim of the waste receptor must be at least twice the pipe's inside diameter. This type of indirect waste connection offers a high level of protection by not allowing any possibility for backsiphoning of waste or sewage.

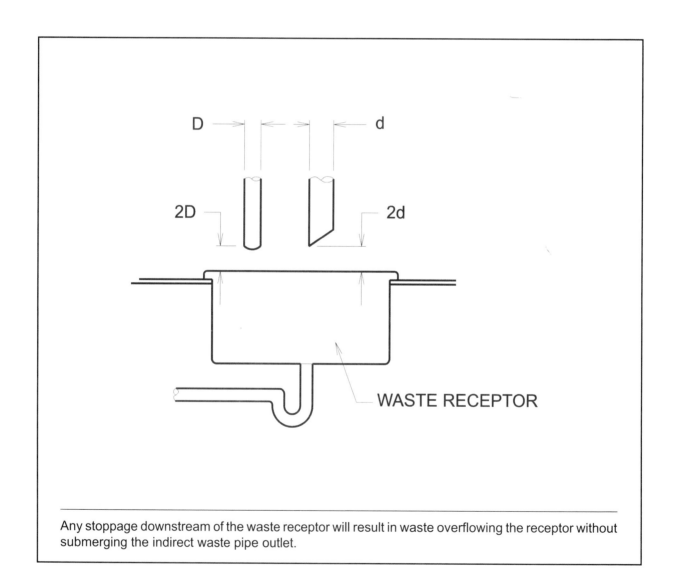

Any stoppage downstream of the waste receptor will result in waste overflowing the receptor without submerging the indirect waste pipe outlet.

Topic: Waste Receptors

Category: Indirect/Special Waste

Reference: IPC 802.3

Subject: Indirect Wastes

Code Text: *Every waste receptor shall be of an approved type. A removable strainer or basket shall cover the waste outlet of waste receptors. Waste receptors shall be installed in ventilated spaces. Waste receptors shall not be installed in bathrooms or toilet rooms or in any inaccessible or unventilated space such as a closet or storeroom. Ready access shall be provided to waste receptors.*

Discussion and Commentary: There are no specific standards regulating waste receptors. They may be identified by various names, including floor drain, open hub drain, floor sink and waste sink. Floor sinks are specialized sinks designed to be installed in the floor. Manufacturers produce a special line of fixtures designed to be used as waste receptors, some of which look like kitchen sinks.

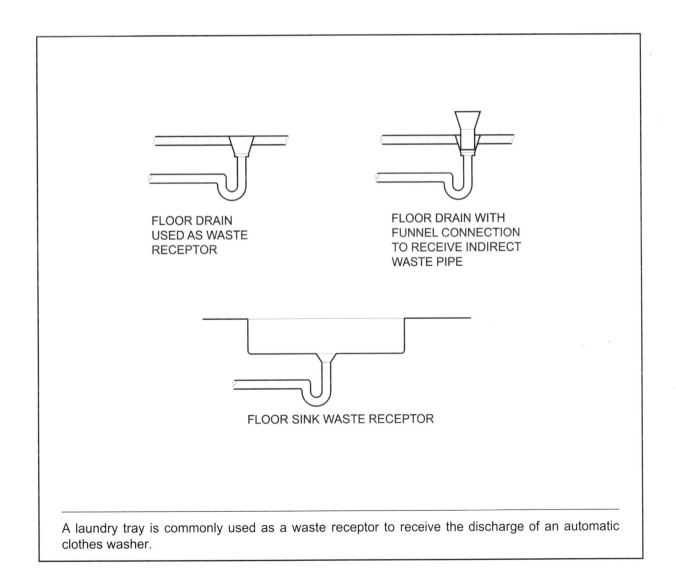

FLOOR DRAIN USED AS WASTE RECEPTOR

FLOOR DRAIN WITH FUNNEL CONNECTION TO RECEIVE INDIRECT WASTE PIPE

FLOOR SINK WASTE RECEPTOR

A laundry tray is commonly used as a waste receptor to receive the discharge of an automatic clothes washer.

Code Text: *A waste receptor shall be sized for the maximum discharge of all indirect waste pipes served by the receptor. Receptors shall be installed to prevent splashing or flooding.*

Discussion and Commentary: Sizing a waste receptor is based on the probability of simultaneous discharge of the various fixtures connecting to the same waste receptor. Two major problems in the design and sizing of a waste receptor are splashing and flooding. Both must be minimized to avoid creating an insanitary condition. Both the shape of the waste receptor and the discharge rate of the indirect waste pipe affect the amount of splashing and flooding involved. The receptor must be capable of accepting all of the discharge without overflowing. While accepting the discharge of an indirect waste pipe, the receptor is simultaneously discharging to the drainage system at a rate based on the size of its drain pipe.

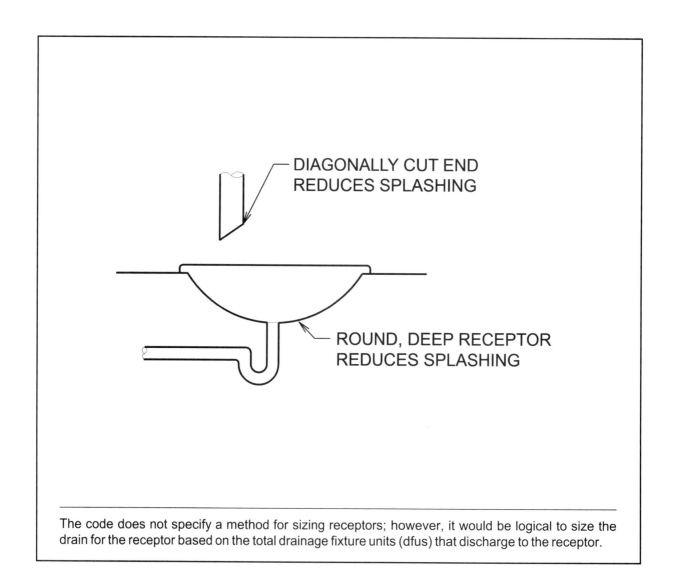

DIAGONALLY CUT END
REDUCES SPLASHING

ROUND, DEEP RECEPTOR
REDUCES SPLASHING

The code does not specify a method for sizing receptors; however, it would be logical to size the drain for the receptor based on the total drainage fixture units (dfus) that discharge to the receptor.

Code Text: *Waste receptors shall be permitted in the form of a hub or pipe extending not less than 1 inch (25.4 mm) above a water-impervious floor and are not required to have a strainer.*

Discussion and Commentary: An open hub drain is a common method of providing a waste receptor in water-impervious floors. It eliminates the need to provide a separate fixture with a strainer. A hubbed drain pipe or plain-end section of pipe serves as the receptor for receiving indirect waste. Because there is no strainer required, an open hub drain is restricted to clear-water waste without solids.

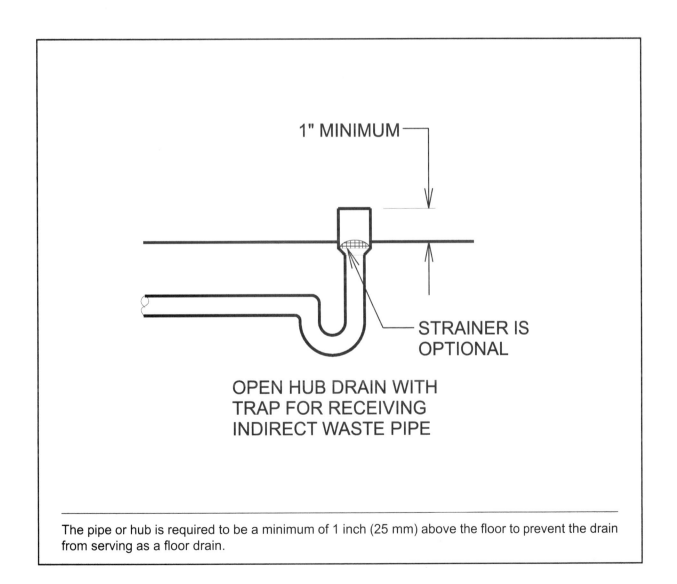

1" MINIMUM

STRAINER IS OPTIONAL

OPEN HUB DRAIN WITH TRAP FOR RECEIVING INDIRECT WASTE PIPE

The pipe or hub is required to be a minimum of 1 inch (25 mm) above the floor to prevent the drain from serving as a floor drain.

Code Text: *Standpipes shall be individually trapped. Standpipes shall extend a minimum of 18 inches (457 mm) and a maximum of 42 inches (1066 mm) above the trap weir. Access shall be provided to all standpipes and drains for rodding.*

Discussion and Commentary: A standpipe can serve as an indirect waste receptor. The minimum dimension for the height of the receptor is intended to provide minimal retention capacity and head pressure to prevent an overflow. This is especially necessary when an indirect waste pipe receives a high rate of discharge, such as the pumped discharge of a clothes washer. The maximum height requirement is intended to control the waste flow velocity at the trap inlet. Excessive inlet velocity can promote trap siphonage.

Trap weir is the bottom surface of the trap outlet pipe that water flows out of. At this elevation, the water surface is at the same level on both sides of the trap.

Topic: Wastewater Temperature

Category: Indirect/Special Waste

Reference: IPC 803.1

Subject: Special Wastes

Code Text: *Steam pipes shall not connect to any part of a drainage or plumbing system and water above 140°F (60°C) shall not be discharged into any part of a drainage system. Such pipes shall discharge into an indirect waste receptor connected to the drainage system.*

Discussion and Commentary: Waste water with a temperature in excess of 140°F (60°C) is considered a special waste because it can soften pipe wall, or erode or damage the drainage piping. This section establishes where special waste systems are required and how they are to be designed.

Dishwashing machines quite often supply hot water at a temperature of 180°F (82°C). Although initially above the threshold temperature of 140°F (60°C), the water is usually reduced after the washing or rinsing cycle to a temperature that will not adversely affect the system piping.

If waste temperatures are too high [above 140°F (60°C)], the waste must be cooled or special piping material must be installed to receive the waste. Waste retention tanks are used as the means of cooling waste.

Code Text: *Corrosive liquids, spent acids or other harmful chemicals that destroy or injure a drain, sewer, soil or waste pipe, or create noxious or toxic fumes or interfere with sewage treatment processes shall not be discharged into the plumbing system without being thoroughly diluted, neutralized or treated by passing through an approved dilution or neutralizing device. Such devices shall be automatically provided with a sufficient supply of diluting water or neutralizing medium so as to make the contents noninjurious before discharge into the drainage system. The nature of the corrosive or harmful waste and the method of its treatment or dilution shall be approved prior to installation.*

Discussion and Commentary: A special waste system must treat any deleterious waste and reduce the threat to an acceptable level before being discharged to the sanitary drainage system. Special waste systems are associated with atypical plumbing systems, including laboratories, chemical plants, processing plants and hospitals. Treatment generally involves dilution of the chemicals waste to an acceptable level.

The code does not specify what is an *acceptable level*, because it varies by pipe material and type of waste.

Topic: System Design

Reference: IPC 803.3

Category: Indirect/Special Waste

Subject: Special Wastes

Code Text: *A chemical drainage and vent system shall be designed and installed in accordance with this code. Chemical drainage and vent systems shall be completely separated from the sanitary systems. Chemical waste shall not discharge to a sanitary drainage system until such waste has been treated in accordance with Section 803.2.*

Discussion and Commentary: When a chemical drainage system is installed, the drainage and vent piping must still conform to all code requirements. The pipe sizing and installation would have to conform to the requirements of Chapters 7 and 9, and other chapters as applicable. Although it is understood that the drainage piping must be separate until the waste is treated or neutralized, it is always questionable as to whether a separate system is necessary for a vent.

The code requires a completely separate vent system because of the uncertainty of the nature of fumes and vapors in a chemical waste system. The gases in the chemical drain, waste and vent (DWV) system could be incompatible with the piping materials and gases in the normal DWV system.

Quiz

Study Session 11
IPC Chapter 8

1. Fixtures _____ are required to discharge through an indirect waste connection to the sanitary system.

    a. utilized for the storage and preparation of food

    b. utilized for cleaning of eating and cooking utensils

    c. that discharge substances harmful to the sewer system

    d. that discharge wastewater with temperatures in excess of 120°F

    Reference _____

2. Which of the following are allowed to discharge through an air break into an indirect waste receptor?

    a. ice storage bin

    b. scullery sink

    c. commercial dishwashing machine

    d. temperature and pressure relief valve drain

    Reference 802.2.7

3. Floor drains in walk-in refrigerators of food establishments shall be _____.

    a. directly connected to the drainage system

    b. directly connected to the drainage system when protected with a backwater valve

    (c.) indirectly connected to the drainage system by means of an air gap

    d. indirectly connected to the drainage system by means of an air break

Reference 802.12

4. Fixtures that discharge through an indirect waste connection are required to be trapped when the fixture _____ .

    a. drain is 2 inches or larger

    b. discharges clear-water waste

    (c.) drain exceeds 2 feet horizontally in developed length

    d. drain discharges through an air break into an indirect waste receptor located in the same room as the fixture

Reference _____

5. The fixture drain from an indirect waste receptor, receiving the discharge from a two-compartment scullery sink with a 2-inch waste, and a single compartment culinary sink with a $1^1/_2$-inch waste, is required to be _____ inches.

    a. $1^1/_2$ (b.) 2

    c. 3   d. 4

Reference _____

6. A hub drain receiving the discharge from an ice storage bin through an air gap is not required to _____ .

    a. be trapped and vented

    (b.) be equipped with a removable strainer

    c. be designed and installed to prevent splashing

    d. extend a minimum of 1 inch above the surrounding floor

Reference _____

7. Chemical waste is prohibited from being discharged to the sanitary drainage system unless it is first _____ .

    a. permitted for discharge by state and federal agencies

    b. filtered through a diatomaceous earth filtration system

    c. diluted by water or passed through a neutralizing medium

    d. tested to determine that it is not corrosive to the piping system

    Reference 803.2

8. The maximum temperature of water allowed to discharge to any part of the drainage system is _____ .

    a. 120°F   b. 140°F

    c. 180°F   d. 210°F

    Reference _____

9. Chemical drainage systems are required to comply with which of the following?

    a. separation from the sanitary drainage and vent system

    b. separation from the sanitary drainage system piping only

    c. restricted to installation in certified laboratories

    d. components must be third party certified

    Reference _____

10. Automatic clothes washer standpipes are limited to a maximum height of _____ inches above the trap weir.

    a. 18   b. 24

    c. 30   d. 42

    Reference _____

11. Indirect waste pipes and systems shall be constructed _____ .

    a. of materials complying with Chapter 7

    b. of third party certified materials

    c. in accordance with Chapter 3

    d. in accordance with the referenced standards

    Reference _____

12. Indirect waste receptors are allowed to be installed in all of the following locations, *except* _____ .

    a. boiler rooms   b. bathrooms

    c. ventilated pantries   d. soiled linen utility rooms

    Reference _____

13. Which of the following statements is *true*?

    a. Swimming pool deck drains are not allowed to discharge to the building drainage system.

    b. Swimming pool backwash water must be neutralized before discharging to the drainage system.

    c. Swimming pool filtration piping must be third party certified as a special waste component.

    d. Swimming pool backwash water, if discharged to a building drain, must be through an indirect waste connection.

    Reference _____

14. All indirect waste piping that exceeds _____ feet in total developed length shall be trapped.

    a. 2   b. 3

    c. 4   d. 5

    Reference _____

15. Standpipes shall extend a minimum of _____ inches and a maximum of _____ inches above the trap.

   a. 6, 24   b. 6, 32

   c. 18, 36   d. 18, 42

   Reference _____

16. A domestic dishwashing machine may discharge into a wye-branch fitting on the tailpiece of the kitchen sink, provided the waste line is _____ .

   a. connected to a deck mounted air break

   b. fastened to the underside of the counter

   c. a minimum 1-inch diameter

   d. trapped

   Reference 802.6

17. A waste receptor shall _____ .

   a. have a removable strainer

   b. be permitted in a bathroom

   c. not be permitted in a ventilated store room

   d. be permitted behind an access panel

   Reference _____

18. Waste receptors shall be permitted in the form of a hub or pipe extending not less than _____ inch(es) above a water-impervious floor.

   a. $^3/_4$ b. 2

   c. $1^1/_2$ d. 1

   Reference _____

19. The air gap between the indirect waste pipe and the flood level rim of the waste receptor shall be a minimum of _____ times the effective opening of the indirect waste pipe.

    a. 1   (b.) 2 Remember.

    c. 3   d. 4

    Reference _____

20. Special function sterilization, distillation, processing, cooling and food storage equipment shall be provided with protection against 801.2 .

    a. splashing   (b.) fouling

    c. damage   d. displacement

    Reference _____

21. When fixtures discharging to the drainage system are not required to be indirectly connected, they _____ .

    (a.) shall be directly connected

    b. are permitted to be directly connected

    c. are permitted to be indirectly connected

    d. may be directly connected if approved

    Reference _____

22. In a walk-in food storage freezer, the waste line serving the floor drain shall 802.1.2

    a. be trapped

    (b.) not be trapped

    c. be protected from freezing

    d. be directly connected to the drainage system

    Reference _____

23. A waste receptor shall be sized for _____ percent of the maximum discharge of all indirect waste pipes served by the receptor.

    a. 50   b. 100

    c. 150   d. 200

Reference 802.31

24. A neutralizing device for corrosive wastes shall make the contents _____ before discharge into the drainage system.

    a. neutral   b. harmless

    c. nontoxic   d. noninjurious

Reference 803.2

25. Steam pipes _____ to a drainage system.

    a. shall not connect

    b. shall discharge indirectly

    c. are permitted to discharge indirectly

    d. are permitted to connect directly

Reference _____

Study Session

2006 IPC Sections 901– 909
Vents I

OBJECTIVE: To gain an understanding of the code requirements related to trap seal protection, common vents, wet venting and vent termination.

REFERENCE: Sections 901 – 909, 2006 *International Plumbing Code*

KEY POINTS:
- Vent piping prevents what from occurring in the system?
- Are all traps and trapped fixtures required to be vented?
- Is it permitted to combine the vent system for a chemical waste system and other venting systems?
- Can the vent system be used for purposes other than the venting of the plumbing?
- What section provides the provisions for testing the venting system?
- What is engineered venting?
- Open vents through roofs are required to extend above the roof how many inches?
- Where frost closure is possible, what is the minimum size of the vent extension through the roof?
- What protection is required to prevent water intrusion when a vent penetrates the roof?
- What minimum distance is required between an open vent terminal in regard to windows, doors and air intakes?
- What dimensions are required for the vent distances above grade and away from property lines?
- Drainage off the vent piping is required to be provided by what means?
- Which types of fixtures are permitted to be wet vented?
- What is horizontal wet venting?
- What is vertical wet venting?

Code Text: *The vent system serving each building drain shall have at least one vent pipe that extends to the outdoors.*

Discussion and Commentary: This is a requirement to equalize pressures in the drainage system on each side of the fixture trap. At least one vent pipe must be extended to the outside to minimize the potential for problems. There are several venting methods that could be used in different situations. Mechanical units such as air admittance valves are also allowed under some circumstances.

The drainage/waste/vent system must always have close to equalized system pressure for proper functioning of the plumbing system.

Topic: Vent Stack Required	**Category:** Vents
Reference: IPC 903.2	**Subject:** Outdoor Vent Extension

Code Text: *A vent stack shall be required for every drainage stack that is five branch intervals or more.*

Discussion and Commentary: If a drainage stack is five or more branch intervals in height, a vent stack is required. If the drainage stack is less than five branch intervals in height, a vent stack is not required because the pressures in the drainage stack are not likely to create a pressure differential at the trap seals in excess of 1 inch of water column (249 Pa). A stack vent is typically utilized as a collection point for vent pipes so that a single roof penetration can be made.

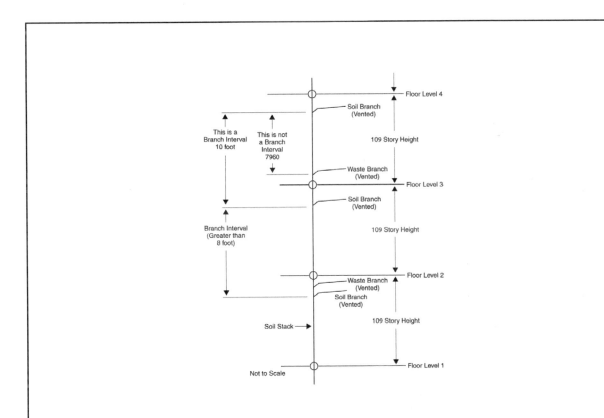

BRANCH INTERVAL: A vertical measurement of distance, 8 feet (2438 mm) or more in developed length, between the connections of horizontal branches to a drainage stack. Measurements are taken down the stack from the highest horizontal branch connection.

VENT STACK: A vertical vent pipe installed primarily for the purpose of providing circulation of air to and from any part of the drainage system.

Code Text: *Every vent stack or stack vent shall terminate outdoors to the open air or to a stack-type air admittance valve in accordance with Section 917.*

Discussion and Commentary: A vent stack or stack vent cannot terminate in an attic, cupola, open parking structure or any other space within a building envelope, regardless of how well such space is ventilated. The vent must extend to the outdoors to permit gases, vapors and odors from the drainage system to escape into the atmosphere away from the building. If air admittance valves are to be used, they must be stack-type air admittance valves.

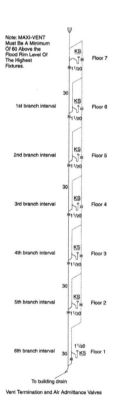

Note: MAXI-VENT Must Be A Minimum Of 60 Above the Flood Rim Level Of The Highest Fixtures.

1st branch interval — KS — Floor 7

30

1st branch interval — KS — Floor 6

2nd branch interval — KS — Floor 5

30

3rd branch interval — KS — Floor 4

4th branch interval — KS — Floor 3

30

5th branch interval — KS — Floor 2

6th branch interval — 30 — KS — Floor 1

To building drain

Vent Termination and Air Admittance Valves

AIR ADMITTANCE VALVE: One-way valve designed to allow air to enter the plumbing drainage system when negative pressures develop in the piping system. The device shall close by gravity and seal the vent terminal at zero differential pressure (no flow conditions) and under positive internal pressures.

Code Text: *Every vent stack shall connect to the base of the drainage stack. The vent stack shall connect at or below the lowest horizontal branch. Where the vent stack connects to the building drain, the connection shall be located downstream of the drainage stack and within a distance of 10 times the diameter of the drainage stack.*

Discussion and Commentary: A drainage stack of five or more branch intervals must be vented at or below the lowest branch connection to relieve the positive pressure developed in the stack. The connection can be to the stack or to the building drain downstream of the stack and within a distance of 10 times the drainage stack diameter.

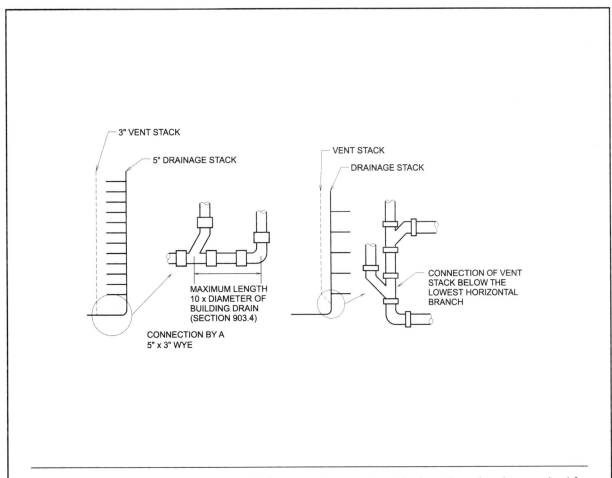

The size of the vent connection to the drainage system must not be less than the size required for the vent stack. Vent stacks are sized in accordance with Section 916.1, based on the size of the drainage stack at its base.

Topic: Roof Extension

Category: Vents

Reference: IPC 904.1

Subject: Vent Terminals

Code Text: *All open vent pipes that extend through a roof shall be terminated at least [NUMBER] inches (mm) above the roof, except that where a roof is to be used for any purpose other than weather protection, the vent extensions shall be run at least 7 feet (2134 mm) above the roof.*

Discussion and Commentary: Where the roof is occupied for any purpose, such as for recreational or assembly purposes, plumbing vents must extend at least 7 feet (2134 mm) above the roof to prevent harmful sewer gases from contaminating the area. The sewer gases will tend to disperse into the air, rather than accumulate near the roof surface. The minimum termination height must be determined by the local jurisdiction based on snowfall rates.

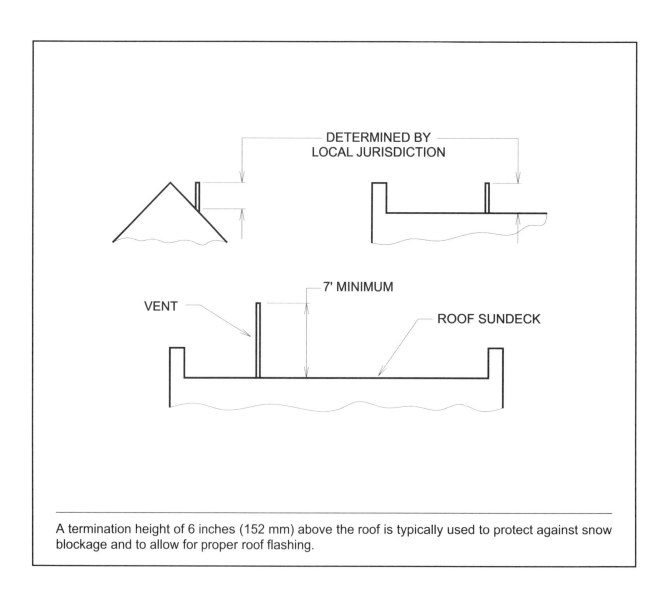

A termination height of 6 inches (152 mm) above the roof is typically used to protect against snow blockage and to allow for proper roof flashing.

Topic: Location of Vent Terminal

Category: Vents

Reference: IPC 904.5

Subject: Vent Terminals

Code Text: *An open vent terminal from a drainage system shall not be located directly beneath any door, openable window, or other air intake opening of the building or of an adjacent building, and any such vent terminal shall not be within 10 feet (3048 mm) horizontally of such an opening unless it is at least 2 feet (610 mm) above the top of such opening.*

Discussion and Commentary: The vent terminal must be located away from any building air intake opening (gravity or mechanical) to reduce the possibility of sewer gases entering a building.

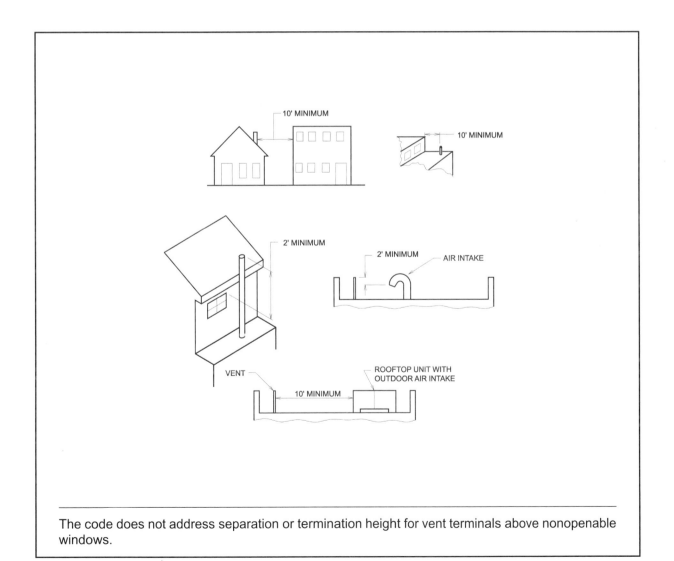

The code does not address separation or termination height for vent terminals above nonopenable windows.

Code Text: *All vent and branch vent pipes shall be so graded and connected as to drain back to the drainage pipe by gravity.*

Discussion and Commentary: Vents convey water vapor from the drainage system and such vapor may condense in the vent piping. Also, rainwater can enter a vent system at the vent termination; therefore, the vent must be sloped to the drainage system, thus preventing any accumulation of condensate or rainwater. The slope of a vent pipe does not affect the movement of air inside the pipe. Vent piping must be adequately supported to maintain slope and prevent sagging.

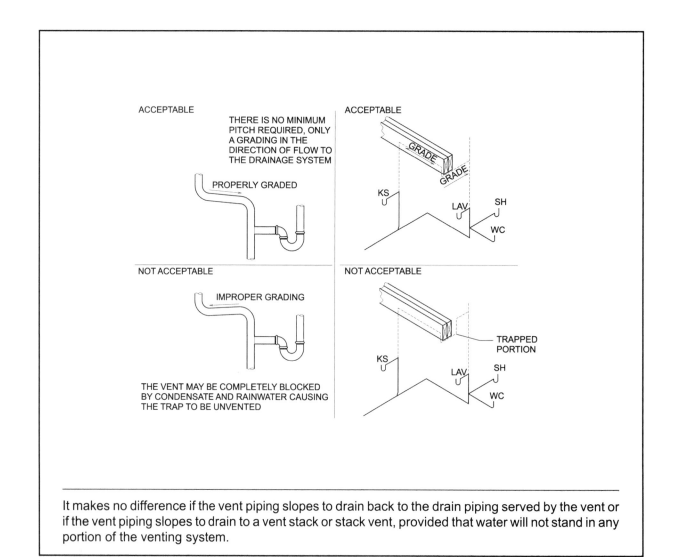

It makes no difference if the vent piping slopes to drain back to the drain piping served by the vent or if the vent piping slopes to drain to a vent stack or stack vent, provided that water will not stand in any portion of the venting system.

Code Text: *Every dry vent connecting to a horizontal drain shall connect above the centerline of the horizontal drain pipe.*

Discussion and Commentary: When drainage enters a pipe, it is naturally assumed that the liquid proceeds down the pipe in the direction of flow. Being a liquid, however, the drainage seeks its own level and may move against the direction of flow before reversing and draining down the pipe. When this occurs, the solids may drop out of suspension to the bottom of the pipe. In a drain, the solids will again move down the pipe with the next discharge of liquid into the drain.

To avoid possible blockages, the vent must connect to horizontal drains above the centerline. The intent is to prevent waste from entering a dry vent by connecting the vent above the flow line of the drain. Such connections will result in vent piping that forms a 45-degree (0.79 rad) angle or greater with the horizontal.

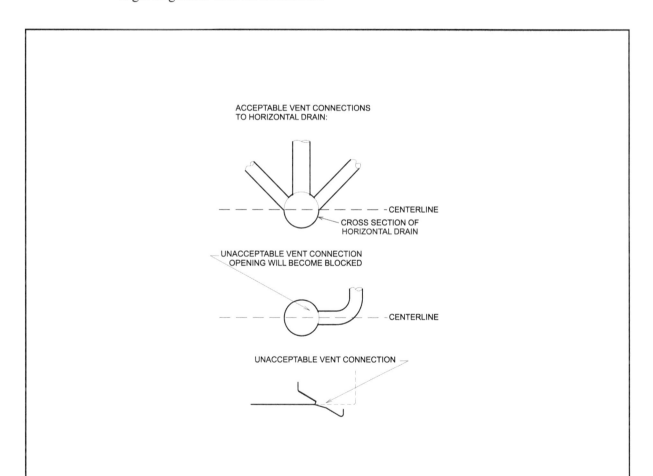

VENT SYSTEM: A pipe or pipes installed to provide a flow of air to or from a drainage system, or to provide a circulation of air within such system to protect trap seals from siphonage and backpressure.

Code Text: *Where the drainage piping has been roughed-in for future fixtures, a rough-in connection for a vent shall be installed. The vent size shall be not less than one-half the diameter of the rough-in drain to be served. The vent rough-in shall connect to the vent system, or shall be vented by other means as provided for in this chapter. The connection shall be identified to indicate that it is a vent.*

Discussion and Commentary: Section 710.2 addresses the installation of drainage pipe for future plumbing fixtures. Where future fixture rough-in piping occurs, vent piping must also be installed. The vent pipe is simply roughed-in in such a manner as to remain accessible for future connection when plumbing fixtures are installed. The vent rough-in must be tied in to the vent system, as required by Section 905.1, or it must extend to a vent terminal in the outside air.

710.2 Future fixtures. Where provision is made for the future installation of fixtures, those provided for shall be considered in determining the required sizes of drain pipes.

Where the installation of a future fixture will require making a connection to the vent rough-in, such connection must be identified as a vent so that the purpose of the original installation is evident when the future fixture is installed.

Code Text: *Each fixture trap shall have a protecting vent located so that the slope and the developed length in the fixture drain from the trap weir to the vent fitting are within the requirements set forth in Table 906.1.*

Exception: The developed length of the fixture drain from the trap weir to the vent fitting for self-siphoning fixtures, such as water closets, shall not be limited.

Discussion and Commentary: The distance from every trap to its vent is limited to reduce the possibility of the trap self-siphoning. Self-siphoning of a trap is the siphoning caused by the discharge from the fixture the trap serves.

Studies have been done to address whether the trap-to-vent distance for a water closet should be limited. Because the water closet relies on self-siphonage to operate properly, and the trap is resealed after each use, it may be argued that there is no need for limiting the distance from the water closet to a vent.

TABLE 906.1
MAXIMUM DISTANCE OF FIXTURE TRAP FROM VENT

SIZE OF TRAP (inches)	SLOPE (inch per foot)	DISTANCE FROM TRAP (feet)
$1\frac{1}{4}$	$\frac{1}{4}$	5
$1\frac{1}{2}$	$\frac{1}{4}$	6
2	$\frac{1}{4}$	8
3	$\frac{1}{8}$	12
4	$\frac{1}{8}$	16

For SI: 1 inch = 25.4 mm, 1 foot = 304.8 mm,
1 inch per foot = 83.3 mm/m.

The water closet must be vented in all cases, but limiting the distance between the vent connection and the water closet appears to serve no purpose.

Code Text: *The total fall in a fixture drain due to pipe slope shall not exceed the diameter of the fixture drain, nor shall the vent connection to a fixture drain, except for water closets, be below the weir of the trap.*

Discussion and Commentary: This section reinforces the previous requirement in Section 906.1. The trap weir must stay below the highest inlet to the vent except for fixtures with integral traps. The fixture drain cannot offset or drop vertically, other than the required slope, between the trap and its vent connection.

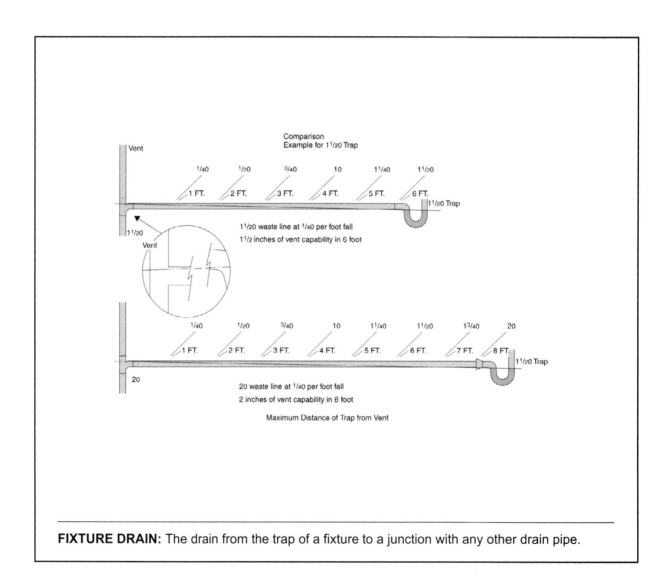

FIXTURE DRAIN: The drain from the trap of a fixture to a junction with any other drain pipe.

Topic: Crown Vent

Category: Vents

Reference: IPC 906.3

Subject: Fixture Vents

Code Text: *A vent shall not be installed within two pipe diameters of the trap weir.*

Discussion and Commentary: Without proper venting distance, the vent opening becomes blocked. The blockage is a result of the action of the drainage flowing through the trap. The flow direction and velocity will force waste up into the vent connection, eventually clogging it with debris. It has been determined that the vent connection must be a minimum of two pipe diameters downstream from the trap weir to prevent the vent from becoming blocked.

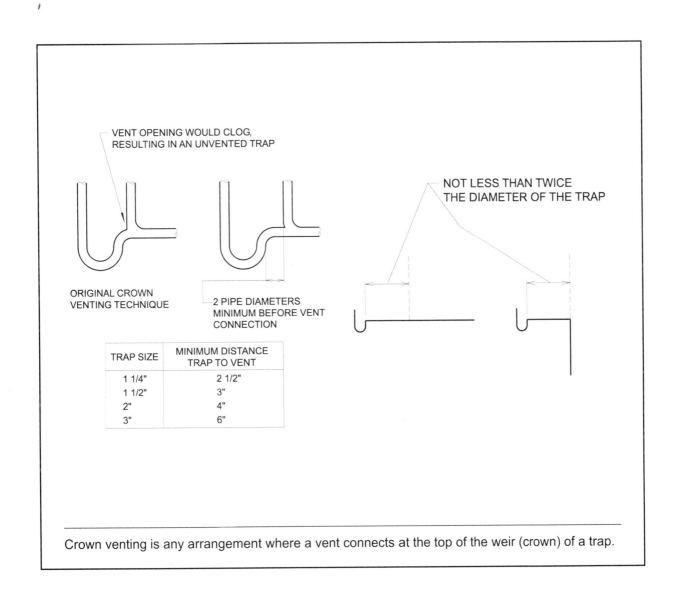

TRAP SIZE	MINIMUM DISTANCE TRAP TO VENT
1 1/4"	2 1/2"
1 1/2"	3"
2"	4"
3"	6"

Crown venting is any arrangement where a vent connects at the top of the weir (crown) of a trap.

Code Text: *An individual vent is permitted to vent two traps or trapped fixtures as a common vent. The traps or trapped fixtures being common vented shall be located on the same floor level.*

Discussion and Commentary: For common venting, the vent is classified and sized as an individual vent. The fixture drains being common vented may connect to either a vertical or horizontal drainage pipe. Any two fixtures can be common vented. The fixtures being common vented must be located on the same floor level so as to control flow velocities by limiting the vertical drop.

Where water closets are common vented, Section 706.3 requires directional drainage fittings to prevent the discharge action of one fixture from interfering with the other fixture or causing a blowback into another fixture.

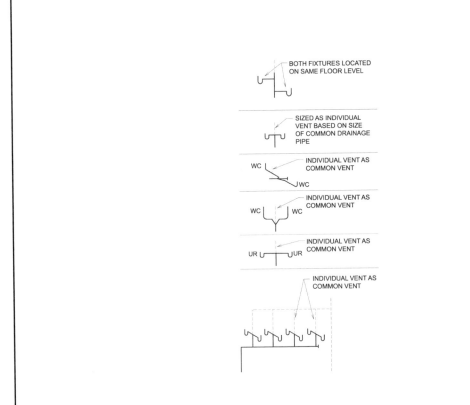

If two water closets are installed back-to-back at the same level and are common vented, the connection to the vertical drainage branch must be through a double combination wye and eighth bend.

Code Text: *The dry-vent connection to the wet vent shall be an individual vent or common vent to the lavatory, bidet, shower or bathtub. In vertical wet-vent systems, the most upstream fixture drain connection shall be a dry-vented fixture drain connection. In horizontal wet-vent systems, not more than one wet-vented fixture drain shall discharge upstream of the dry-vented fixture drain connection.*

Discussion and Commentary: Vertical venting and horizontal wet venting is similar to combination drain and vent systems. The wet vented section of piping is enlarged to handle the waste and allow additional free area in the pipe to serve as a vent for a downstream fixture.

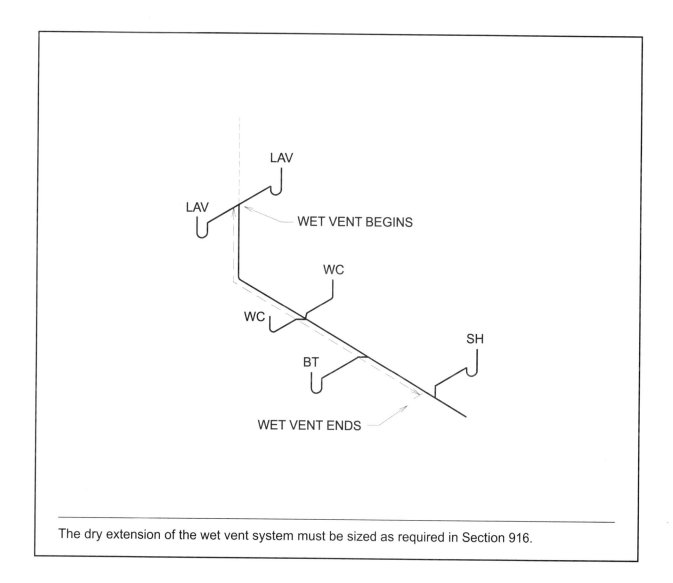

The dry extension of the wet vent system must be sized as required in Section 916.

Code Text: *The dry vent serving the wet vent shall be sized based on the largest required diameter of pipe within the wet vent system served by the dry vent. The wet vent shall be of a minimum size as specified in Table 909.3, based on the fixture unit discharge to the wet vent.*

Discussion and Commentary: To ensure proper amount of air flow for the entire system, the size of the dry vent serving the wet vent is based on the largest pipe diameter within the wet vent system. This is the same criteria used for both systems. Table 909.3 establishes prescriptive wet vent pipe systems for different fixture unit loads from 1 dfu (a single lavatory or bidet) to 12 dfus (two bathroom groups).

TABLE 909.3
WET VENT SIZE

WET VENT PIPE SIZE (inches)	DRAINAGE FIXTURE UNIT LOAD (dfu)
$1^1/_2$	1
2	4
$2^1/_2$	6
3	12

For SI: 1 inch = 25.4 mm.

In a wet vented system, a single vent can provide airflow when fixtures are being discharged and relieve pressures that develop in the drain piping.

Quiz

Study Session 12
IPC Sections 901 – 909

1. Trap seal protection depends on maintaining air flow downstream from the fixture trap by providing adequate _____ .

    a. fixture clearances

    b. fixture venting

    c. protection against low flow velocities in the drainage system

    d. protection against excessive flow velocities in the drainage system

    Reference _____

2. What is the maximum pneumatic pressure differential that is allowed against a trap seal?

    a. 1 inch of water column   b. 1 foot of head

    c. 1 PSI   d. 0.433 PSI

    Reference _____

3. An individual vent for a fixture receiving the discharge of a chemical waste is allowed to terminate to which of the following locations?

    a. to the vent stack for the sanitary drainage system

    b. to a chemical vent hood that vents to the open air, outside the building

    c. to a vent stack or stack vent, independent of the sanitary drainage system

    d. to an air admittance valve connected to the nonneutralized piping system

    Reference _____

4. Which of the following complies with minimum standards for vent pipe flashings?

    a. flashings that are field fabricated from sheet lead of 2.5 lb/sq ft

    b. prefabricated lead flashings that are 2.5 lb/sq ft material weight

    c. prefabricated copper flashings, weighing 4 ounces per square foot

    d. 8-ounce copper flashing, with material in compliance with ASTM B687-99

    Reference _____

5. In a sanitary drainage system, what is the minimum number of vent terminals that are required to extend to the outdoors?

    a. one

    b. two

    c. one for each air admittance valve used in the system

    d. one for each vent stack or stack vent

    Reference _____

6. A vent connection to the base of a stack is required to _____ .

    a. connect above the lowest horizontal branch

    b. connect to the building drain

    c. connect at or below the lowest horizontal branch

    d. connect a minimum of 10 pipe diameters away from the stack

    Reference _____

7. All through-the-roof vent terminals in Alamosa, Colorado, are required to be protected against frost closure by a minimum _____ -inch diameter vent pipe, beginning _____ inches below the roof.

    a. 4, 12   b. 3, 12

    c. 3, 18   d. 4, 6

    Reference _____

8. Which one of the following locations is required to provide frost closure protection for vent terminals?

    a. Denver, Colorado   b. Juneau, Alaska

    c. Goodland, Kansas   d. South Bend, Indiana

Reference _____

9. What is the minimum required height of a vent terminal located through a roof that is used for purposes besides weather protection?

    a. 1 foot   b. 2 feet

    c. 7 feet   d. 10 feet

Reference _____

10. A vent shall terminate a minimum of _____ feet horizontally from or at least _____ feet above a window opening.

    a. 8, 2   b. 10, 2

    c. 8, 4   d. 10, 4

Reference _____

11. What is the minimum required height above average grade for a vent terminating through the side wall of a building?

    a. 2 feet   b. 7 feet

    c. 8 feet   d. 10 feet

Reference _____

12. Which of the following locations is required to provide protection against freezing by providing insulation, heat or both for vent piping installed on the exterior of the building?

    a. Chicago, Illinois, Midway Airport

    b. Goodland, Kansas

    c. Toledo, Ohio

    d. Portland, Maine

Reference _____

13. A vent is prohibited from offsetting horizontally until the vertical rise of the vent is
_____ .

    a. above the highest fixture branch connected to the building drain

    b. above the flood level rim of the highest fixture connecting to the building drain

    c. 6 inches above the connection of the stack vent to a drainage stack

    d. 6 inches above the flood rim of the highest trap or trapped fixture being vented

Reference _____

14. A kitchen sink with a $1^1/_2$-inch fixture drain must be vented within a developed length of
_____ feet from the trap weir to the vent fitting.

    a. 3   b. 5

    c. 6   d. 8

Reference _____

15. Which fixture is allowed to have a vent connection to the fixture drain below the weir of
the trap?

    a. shower drain   b. laundry sink

    c. floor drain   d. water closet

Reference _____

16. What is the maximum developed length of the fixture drain from the trap weir to the vent
fitting for a water closet?

    a. 8 feet   b. 12 feet

    c. 16 feet   d. unlimited

Reference 906.1

17. How many fixtures are vented by a common vent?

    a. 1   b. 2

    c. 3   d. 4

Reference 908.1

18. The vertical pipe, which connects two fixture drains, one from a water closet and the other from a tub, connecting at different levels and vented by a *common vent* pipe, is sized based on ___B___ .

   a. the upper fixture being the water closet

   b. the maximum discharge from the upper fixture drain

   c. the maximum discharge from the lower fixture drain

   d. one-half the size of the fixture branch receiving the discharge from both fixture drains

   Reference 908.3

19. When a fixture is individually vented, the vent connects to _____ .

   a. a drain receiving the discharge from more than one fixture drain

   b. the fixture branch receiving the discharge

   c. the fixture drain of the trapped fixture

   d. a minimum $1^1/_2$-inch drain

   Reference _____

20. Wet venting is permitted for any combination of fixtures _____ .

   a. within two bathroom groups located on adjacent floors

   b. within two bathroom groups located on the same floor

   c. except water closets and urinals

   d. located on the same floor

   Reference _____

21. The drain of two back-to-back lavatories serves as the wet vent for a combination tub/shower. The maximum length of the tub/shower fixture drain is measured from the _____ .

   a. trap to the connection with the branch drain of the lavatories

   b. trap to the connection with the fixture drain of the lavatories

   c. trap to the connection with the common dry vent

   d. connection with the branch drain to the common dry vent

   Reference _____

22. Fixtures that are not within a bathroom group and discharge to a wet-vented horizontal branch drain are required to be _____ .

    a. circuit vented and connect upstream from the horizontal wet vent

    b. vented separately and connect downstream of the horizontal wet vent

    c. vented by a branch-type air admittance valve conforming to ASSE 1050

    d. vented by a branch-type air admittance valve conforming to ASSE 1051

Reference _____

23. In a horizontal wet-vent system, how many wet-vented fixture drains are allowed to discharge upstream of the dry-vented fixture drain connection?

    a. none   b. one

    c. two   d. three

Reference 909.2

24. The maximum distance of a 2-inch trap on a 2-inch fixture drain from the vent is _____ feet _____ inches.

    a. 2, 6   b. 3, 6

    c. 5, 0   d. 8, 0

Reference _____

25. A vent shall not be installed within _____ pipe diameters of the trap weir.

    a. 2   b. 3

    c. 4   d. 6

Reference 906.

Study Session

13

2006 IPC Sections 910 – 919
Vents II

OBJECTIVE: To develop an understanding of the code provisions for vents, circuit venting, combination drain and vent systems, island fixture venting and the uses of air admittance valves.

REFERENCE: Sections 910 – 919, 2006 *International Plumbing Code*

KEY POINTS:
- A vent stack is required based on how many branch intervals?
- What is a combination waste and vent system?
- When a common vent header is used, what determines the size?
- Dry vents connect at what point to a horizontal drainpipe?
- What is the minimum height above the flood level rim for vertical rise of the vent?
- What table applies to the distances from the trap weir to the vent opening?
- What determines the size of the waste stack?
- When is a relief vent required?
- What limitations are placed on a combination drain and vent system?
- Which types of vents are permitted to terminate to an air admittance valve?
- What is the minimum height above the fixture drain for the placement of the air admittance valve?
- Why is the air admittance valve required to be located in a ventilated space?

Code Text: *A waste stack shall be considered a vent for all of the fixtures discharging to the stack where installed in accordance with the requirements of this section.*

Discussion and Commentary: A waste stack vent uses the waste stack as the vent for fixtures other than urinals and water closets. The principles of use are based on some of the original research that was done in plumbing. The system has been identified by a variety of names, including vertical wet vent, Philadelphia single-stack and multifloor stack venting.

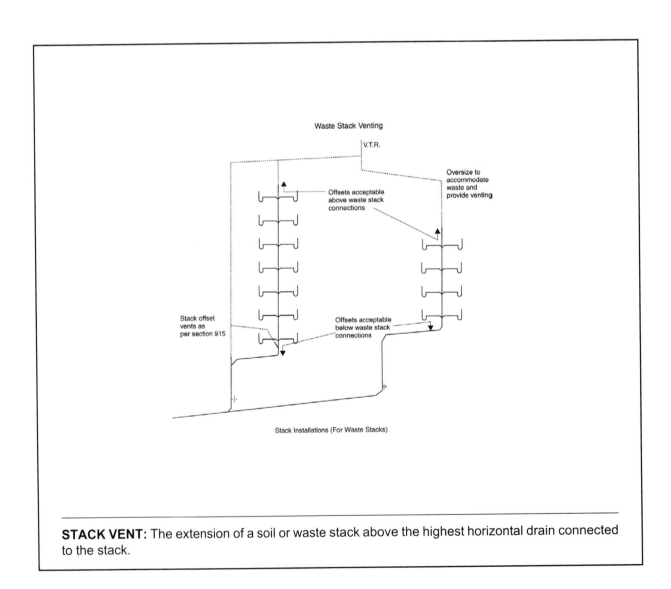

STACK VENT: The extension of a soil or waste stack above the highest horizontal drain connected to the stack.

Code Text: *The waste stack shall be vertical, and both horizontal and vertical offsets shall be prohibited between the lowest fixture drain connection and the highest fixture drain connection. Every fixture drain shall connect separately to the waste stack. The stack shall not receive the discharge of water closets or urinals.*

Discussion and Commentary: Because the drainage stack serves as the vent, there are certain limitations on the design to prevent pressures in the system from exceeding plus or minus 1 inch of water column (249 Pa). Water closets and some urinals have a surging discharge that results in too great a pressure fluctuation in the system, so they are not allowed to connect to the stack system. The system may serve sinks, lavatories, bathtubs, bidets, showers, floor drains, drinking fountains and standpipes.

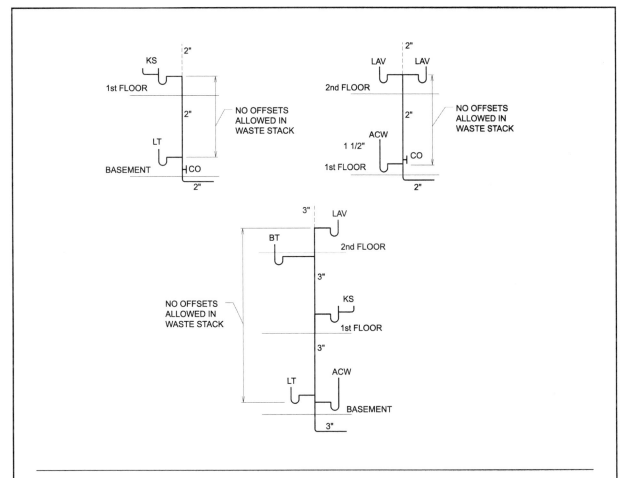

To preserve the desirable annular flow in the stack, offsets (of any degree) are prohibited. The stack must be vertical [90 degrees (1.58 rad) from horizontal] from the lowest fixture connection to above the highest fixture connection.

Code Text: *A stack vent shall be provided for the waste stack. The size of the stack vent shall be not less than the size of the waste stack. Offsets shall be permitted in the stack vent and shall be located at least 6 inches (152 mm) above the flood level of the highest fixture, and shall be in accordance with Section 905.2. The stack vent shall be permitted to connect with other stack vents and vent stacks in accordance with Section 903.5.*

Discussion and Commentary: A full-size stack vent provides the vent opening to the outdoors, allowing the stack to remain at neutral pressures. Offsets are allowed in the stack vent because such offsets are dry and have no effect on the annular flow in the waste stack.

903.5 Vent headers. Stack vents and vent stacks connected into a common vent header at the top of the stacks and extending to the open air at one point shall be sized in accordance with the requirements of Section 916.1. The number of fixture units shall be the sum of all fixture units on all stacks connected thereto, and the developed length shall be the longest vent length from the intersection at the base of the most distant stack to the vent terminal in the open air, as a direct extension of one stack.

Stack dry vents must not run horizontally below the flood level rim of fixtures, because this may allow waste and solids to enter the vent and impair its function.

Topic: Vent Connection

Category: Vents

Reference: IPC 911.2

Subject: Circuit Venting

Code Text: *The circuit vent connection shall be located between the two most upstream fixture drains. The vent shall connect to the horizontal branch and shall be installed in accordance with Section 905. The circuit vent pipe shall not receive the discharge of any soil or waste.*

Discussion and Commentary: The circuit vent must connect between the two most upstream fixture drains to allow proper circulation of air. The vent is located so that the most upstream fixture discharges past the vent connection, thereby washing that section of pipe and preventing the buildup of solids. For this reason, the most upstream fixture should not be a fixture that is seldom used (such as a floor drain). When back-to-back fixtures are installed, the vent is connected between the last two groups.

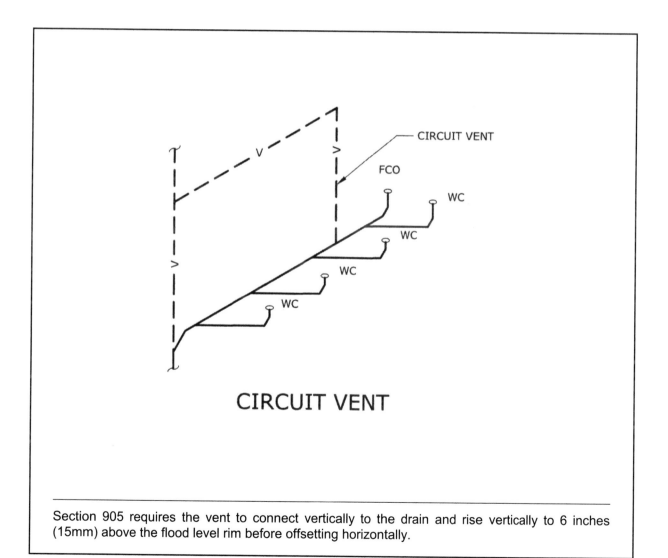

CIRCUIT VENT

Section 905 requires the vent to connect vertically to the drain and rise vertically to 6 inches (15mm) above the flood level rim before offsetting horizontally.

Code Text: *The maximum slope of the vent section of the horizontal branch drain shall be one unit vertical in 12 units horizontal (8-percent slope). The entire length of the vent section of the horizontal branch drain shall be sized for the total drainage discharge to the branch.*

Discussion and Commentary: The principle of a circuit vent is that the drainage branch is entirely horizontal. A maximum pitch of 1 to 12 is to keep the branch horizontal without any vertical offsets.

The horizontal drainage branch must also be uniformly sized to establish consistent flow characteristics and the free movement of air. Because it is uniformly sized for its full length, the horizontal branch will be oversized for all but the most downstream portion. Table 710.1(2) is used to size the horizontal branch.

Circuit venting involves up to eight fixtures and a single vent. A dry vent connects the two most upstream fixtures on the horizontal branch.

Code Text: *A relief vent shall be provided for circuit vented horizontal branches receiving the discharge of four or more water closets and connecting to a drainage stack that receives the discharge of soil or waste from upper horizontal branches.*

Discussion and Commentary: A relief vent is required under the following conditions: 1) If a circuit-vented horizontal branch connects to a drainage stack that is receiving drainage discharge from floors above, and 2) if more than three water closets connect to the circuit-vented horizontal branch. The relief vent will prevent any pressure differential in the drainage stack from affecting the horizontal branch by relieving the pressures.

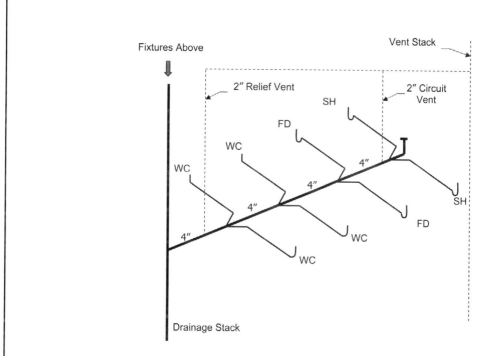

CIRCUIT VENT: A vent that connects to a horizontal drainage branch and vents two traps to a maximum of eight traps or trapped fixtures connected into a battery.

RELIEF VENT: A vent whose primary function is to provide circulation of air between drainage and vent systems.

Topic: Limitation

Category: Vents

Reference: IPC 913.1

Subject: Island Fixture Venting

Code Text: *Island fixture venting shall not be permitted for fixtures other than sinks and lavatories. Residential kitchen sinks with a dishwasher waste connection, a food waste grinder, or both, in combination with the kitchen sink waste, shall be permitted to be vented in accordance with this section.*

Discussion and Commentary: Island fixture venting is a method of venting island sinks and lavatories. Other options such as a combination drain and vent regulated by Section 912 and air admittance valves in accordance with Section 917 are also available. Residential type sinks with dishwasher and/or food waste grinder connections are allowed to be served by an island vent even if not located in a residential occupancy.

Island fixture venting is an option for plumbing fixtures located such that an individual vent cannot be installed without horizontal sections of piping below the flood level rim of the fixture served.

Code Text: *Soil and waste stacks in buildings having more than 10 branch intervals shall be provided with a relief vent at each tenth interval installed, beginning with the top floor.*

Discussion and Commentary: This section requires a method of relieving pressure conditions in drainage stacks more than 10 branch intervals in height. The flow in a drainage stack creates both negative and positive pressures. The vent system is designed to equalize the air pressures at the trap seal. A relief vent is required to assist the fixture venting and the venting at the base of the stack by providing midpoint connections to the stack. The relief vent prevents excessive pressure from being created by emitting or admitting air at specified points within the stack.

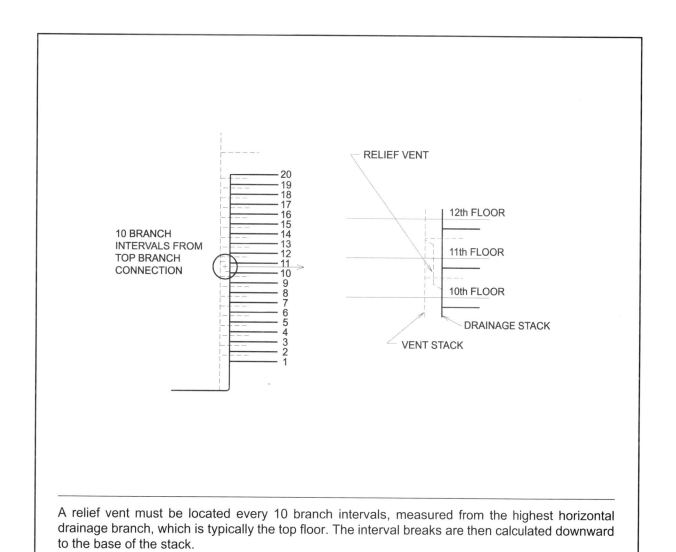

A relief vent must be located every 10 branch intervals, measured from the highest horizontal drainage branch, which is typically the top floor. The interval breaks are then calculated downward to the base of the stack.

Code Text: *The size of the relief vent shall be equal to the size of the vent stack to which it connects. The lower end of each relief vent shall connect to the soil or waste stack through a wye below the horizontal branch serving the floor, and the upper end shall connect to the vent stack through a wye not less than 3 feet (914 mm) above the floor.*

Discussion and Commentary: This section requires the size of the relief vent to be at least equal to the size of the vent stack to which it connects. This is considered the minimum necessary to provide sufficient venting capacity in both the relief vent and the stack vent to which it is connected.

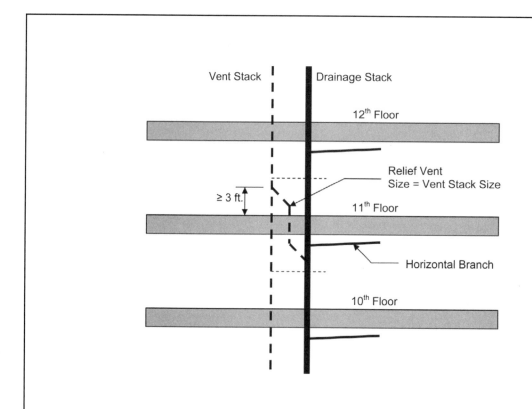

RELIEF VENT: A vent whose primary function is to provide circulation of air between drainage and vent systems.

STACK: A general term for any vertical line of soil, waste, vent or inside conductor piping that extends through at least one story with or without offsets.

SOIL PIPE: A pipe that conveys sewage containing fecal matter to the building drain or building sewer.

Code Text: *Individual, branch and circuit vents shall be permitted to terminate with a connection to an individual or branch-type air admittance valve. Stack vents and vent stacks shall be permitted to terminate to stack-type air admittance valves. Individual and branch-type air admittance valves shall vent only fixtures that are on the same floor level and connect to a horizontal branch drain. The horizontal branch drain having individual and branch-type air admittance valves shall conform to Section 917.3.1 or 917.3.2. Stack-type air admittance valves shall conform to Section 917.3.3.*

Discussion and Commentary: Air admittance valves are allowed to be used in a plumbing system, with certain limitations. Such valves can be used to relieve pressure in individual, branch and circuit vents as well as in stack vents and vent stacks as long as the correct type of air admittance valve is used. ASSE 1050 is the applicable standard for stack-type air admittance valves.

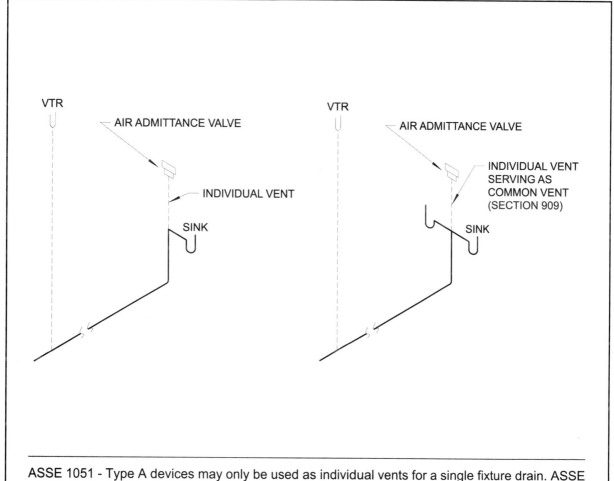

ASSE 1051 - Type A devices may only be used as individual vents for a single fixture drain. ASSE 1051 - Type B devices may serve as branch vents for one or more individual vents, including the dry vent connecting to a common vent, wet vent or circuit vent.

Topic: Stack

Reference: IPC 917.3.3

Category: Vents

Subject: Air Admittance Valves

Code Text: *Stack-type air admittance valves shall not serve as the vent terminal for vent stacks or stack vents that serve drainage stacks exceeding six branch intervals.*

Discussion and Commentary: This text places a restriction on air admittance valves to stacks that are 6 branch intervals or 7 stories tall. Air admittance valves in stack applications function the same as they function in branch applications. The only difference between branch and stack venting is the location of the intervals.

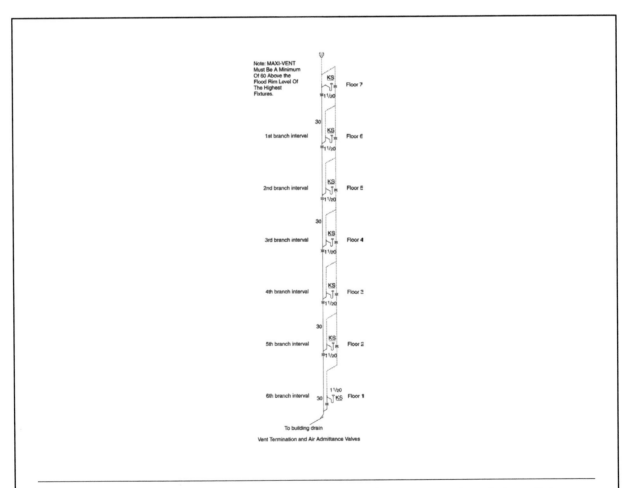

There are benefits to stack venting versus branch venting. In some installations it is more economical because it allows for venting of fixtures on multiple floors, thus reducing the number of valves required.

Topic: Location

Reference: IPC 917.4

Category: Vents

Subject: Air Admittance Valves

Code Text: *Individual and branch-type air admittance valve shall be located a minimum of 4 inches (102 mm) above the horizontal branch drain or fixture drain being vented. Stack-type air admittance valves shall be located not less than 6 inches (152 mm) above the flood level rim of the highest fixture being vented. The air admittance valve shall be located within the maximum developed length permitted for the vent. The air admittance valve shall be installed a minimum of 6 inches (152 mm) above insulation materials.*

Discussion and Commentary: An air admittance valve has one moving part (a seal), which must be maintained a safe distance above the drain served. In the event of a drain stoppage, the seal may become inoperable or operate improperly if waste is permitted to rise into the air admittance valve assembly.

The maximum developed length of a vent pipe equipped with an air admittance valve is regulated by Sections 916.2 and 916.3. Section 916.2 requires a vent exceeding 40 feet (12 192 mm) in developed length (measured from the air admittance valve to the point of connection to the drain pipe) to be increased by one nominal pipe size.

An air admittance valve must be located a safe distance above insulation materials that may block air inlets or otherwise impair the operation of the device.

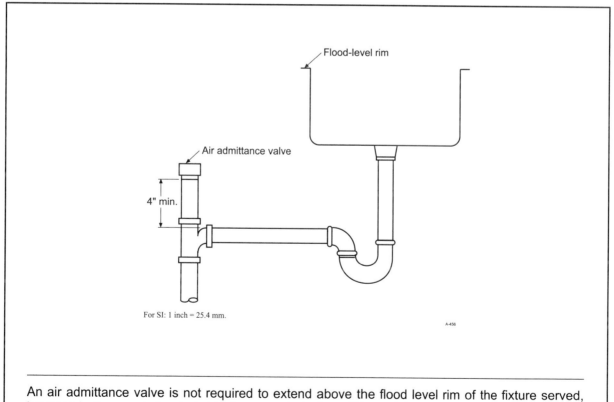

Flood-level rim

Air admittance valve

4" min.

For SI: 1 inch = 25.4 mm.

A-456

An air admittance valve is not required to extend above the flood level rim of the fixture served, because in the event of a drain blockage, such device will trap air between it and the rising waste, thereby protecting the device from contamination.

Quiz

Study Session 13
IPC Sections 910 – 919

1. What is the minimum size required for a stack vent serving a $2^1/_2$-inch waste stack?

   a. $1^1/_4$ inches

   b. $1^1/_2$ inches

   c. 2 inches

   d. $2^1/_2$ inches

   Reference _____

2. What is the minimum size required for a waste stack receiving the discharge of two drainage fixture units at each of three branch intervals?

   a. $1^1/_4$ inches

   b. $1^1/_2$ inches

   c. 2 inches

   d. $2^1/_2$ inches

   Reference _____

3. What is the maximum number of drainage fixture units that is allowed to discharge to a 3-inch diameter waste stack in any one branch interval?

   a. 2

   b. 4

   c. 8

   d. unlimited

   Reference _____

4. Which of the following statements regarding a waste stack vented system is FALSE?

    a. Each of the fixture drains connecting to the stack must connect separately to the stack.

    b. All fixtures that discharge at a high rate of flow are prohibited from connecting to the system.

    c. The fixtures that are allowed to connect to a waste stack system are restricted to those that discharge *waste* only.

    d. The waste stack is sized based on the maximum discharge to the entire stack, and the same size is maintained throughout the length of the stack.

    Reference _____

5. The vent or branch vent for multiple island fixture vents shall extend to a minimum of _____ inches above the highest island fixture being vented before connecting to the outside vent terminal.

    a. 3   b. 4

    c. 6   d. 12

    Reference _____

6. In no case shall the diameter of stack vents and vent stacks be less than one-half the diameter of the drain served or less than _____ inches.

    a. $1^1/_4$ b. $1^1/_2$

    c. 2   d. 3

    Reference _____

7. Where multiple branch vents are connected to a common branch vent, branch vents exceeding _____ feet in developed length shall be increased by one nominal size for the entire developed length of the vent pipe.

    a. 25   b. 30

    c. 35   d. 40

    Reference _____

8. When is a relief vent required to be installed on a circuit-vented branch?

    a. When the circuit-vented branch receives the discharge of four or more water closets and the branch connects to a drainage stack receiving drainage from above.

    b. When the circuit-vented branch is one of multiple circuit-vented branches within a single branch interval.

    c. When the circuit-vented branch has the maximum number of fixture drains connected to the branch.

    d. When the circuit-vented branch receives the discharge of more than 16 dfus and the branch drains are all less than 3-inch diameter fixture drains.

Reference _____

9. The circuit vent is prohibited from receiving any drainage, but a relief vent is permitted to be a fixture drain under what conditions?

    a. when the relief vent receives a maximum discharge of 2 dfus

    b. when the relief vent receives a maximum discharge of 4 dfus

    c. when the relief vent connects to a 4-inch or larger circuit-vented branch

    d. when the relief vent connects to a branch with a 2-inch circuit vent

Reference _____

10. Which of the following fixtures is allowed to be vented by a combination drain and vent system?

    a. floor drain

    b. floor sink receiving the discharge of a clinical sink

    c. urinal

    d. kitchen sink with a food waste grinder

Reference _____

11. What is the maximum allowable slope on a horizontal combination drain and vent pipe?

    a. $\frac{1}{8}$-inch per foot   b. $\frac{1}{4}$-inch per foot

    c. $\frac{1}{2}$-inch per foot   d. 1-inch per foot

Reference _____

12. Where is vertical piping allowed in a combination drain and vent system?

    a. between the fixture outlet and the fixture trap

    b. between the connection to the stack and the building drain

    c. between the fixture drain and the horizontal combination drain and vent pipe

    d. between the connection of the combination drain and vent to the building drain

Reference _____

13. What is the maximum length of a fixture drain for a lavatory connection to the vertical section of a combination drain and vent system?

    a. 2 feet   b. 3 feet

    c. 5 feet   d. 6 feet

Reference _____

14. What is the minimum required size of the combination drain and vent pipe receiving the drainage from two lavatories and a drinking fountain?

    a. $1^1/_2$-inch   b. 2-inch

    c. 3-inch   d. 4-inch

Reference _____

15. All of the following combinations of fixtures are allowed to be vented by an island fixture venting method, *except* _____ .

    a. a residential bar sink with a dishwasher connection

    b. a residential laundry sink and automatic clothes washer

    c. a residential kitchen sink with a dishwasher connection

    d. a residential kitchen sink with a food waste grinder

Reference _____

16. What is the minimum required size of an island fixture vent serving a kitchen sink without a food waste grinder or dishwasher connection?

    a. 1 inch
    b. $1^1/_4$ inches
    c. $1^1/_2$ inches
    d. 2 inches

Reference _____

17. All of the following venting methods are approved for venting the fixture drain from a water closet, *except* _____ .

    a. air admittance valves
    b. circuit vented systems
    c. wet venting systems
    d. combination drain and vent systems

Reference _____

18. Where is a relief vent for a waste stack required?

    a. where buildings have five or more branch intervals

    b. where circuit vented branches connect to a stack

    c. every 10th floor, in buildings with stacks of 10 stories or greater height

    d. at each 10th interval, on stacks having more than 10 branch intervals

Reference _____

19. Which of the following is *true* of relief vents for soil stacks?

    a. The relief vent is required to be 2 inches in all cases.

    b. The relief vent is required to be the same size as the stack.

    c. The relief vent is required to be a minimum of one-half the diameter of the stack.

    d. The relief vent is sized on developed length and the total dfu load on the stack.

Reference _____

20. How is a horizontal offset in a drainage stack in a 20-story building, where the offset is located at the fifth branch interval, required to be vented?

    a. by a vent stack connected to the building drain

    b. by a vent that is equal in size to the drainage pipe offset

    c. by a vent pipe size required for the total dfu discharging to the entire stack

    d. by venting the upper section of the stack, with the offset as the base of the stack

Reference _____

21. Vent pipes exceeding 40 feet in developed length, are required to be _____ .

    a.   sized to a minimum of $1^{1}/_{4}$ inches

    b.   sized to one-half the size of the drain served

    c.   sized equal to the size of the largest drain served

    d.   sized one size greater than the otherwise required size

Reference _____

22. A sump vent for a sewage ejector pump with a discharge capacity of 20 gpm is allowed to be _____ .

    a. $1^{1}/_{4}$ inches for an unlimited length

    b. $1^{1}/_{4}$ inches for a maximum developed length of 40 feet

    c. $1^{1}/_{2}$ inches for an unlimited length

    d. $1^{1}/_{2}$ inches for the first developed length of 40 feet

Reference _____

23. All of the following locations are restricted against use of an air admittance valve, *except* _____ .

    a.   the attic of a residence

    b.   nonneutralized chemical waste systems

    c.   spaces utilized as supply or return air plenums

    d.   termination of drainage stacks exceeding six branch intervals

Reference _____

24. Which of the following fixtures is allowed to connect to a waste stack vent system?

    a.   urinal   b.   water closet

    c.   bedpan washer   d.   automatic clothes washer

Reference _____

25. An air admittance valve shall be installed a minimum of _____ inches above insulation materials.

    a. 8   b. 6

    c. 4   d. 2

Reference _____

Study Session

2006 IPC Chapter 10
Traps, Interceptors and Separators

OBJECTIVE: To develop an understanding of the overall code provisions that regulate the materials, design and installation of sanitary drainage. To develop an understanding of the specific provisions of the code that apply to joints and fittings in the sanitary drainage system.

REFERENCE: Chapter 10, 2006 *International Plumbing Code*

KEY POINTS:
- Generally, every plumbing fixture shall be separated from the drainage system by what type of trap?
- Where is the preferred location of the trap?
- What is the maximum vertical and horizontal distance from the fixture outlet to the trap weir?
- Does the code allow a fixture to be double trapped? ᵘˢᵗ
- What are the characteristics of traps?
- Where are slip joints permitted, and what type of seal is required?
- What are the minimum and the maximum inches of the liquid seal for a fixture trap?
- When would a deeper seal from fixture tapes be allowed?
- Can a trap be larger in diameter than the drainage pipe?
- When are house (building) traps permitted?
- When house traps are installed, what additional provisions apply?
- What is the required size of the relief vent?
- What type of protection is required at the exterior termination of the relief vent?
- What type of protection is required for an underground acid-resisting trap?
- What specific provision applies to pipes and traps located in mental health centers?
- What types of materials create a need for an interceptor?
- How are the size, types, location, design and installation of interceptors and separators determined?
- Where is a grease trap or grease interceptor not required?
- What determines the grease retention capacity?
- The water flow is not permitted to exceed what rate?

KEY POINTS:
(Cont'd)

- Are both treated and untreated light and heavy liquids required to be separated?
- What is the minimum depth of the water seal?
- Where automobiles are serviced, what is the minimum cubic capacity of oil separators for the first 100 square feet of garage floor area?
- What type of access is required for a separator of heavy solids?
- Interceptors for laundries shall have what type of device installed to prevent passage of solids into the system?
- Why are bottling plants required to have an interceptor?
- What types of materials are required to be prevented from entering the drainage system in a slaughterhouse?
- Whenever tight covers are used, what does the design have to take into consideration?
- When interceptors or separators are subject to trap seal loss, what has to be done?
- Preventing the retarding or obstruction of flow in a pipe is the function of what type of fittings?

Topic: Fixture Traps

Reference: IPC 1002.1

Category: Traps, Interceptors and Separators

Subject: Trap Requirements

Code Text: *Each plumbing fixture shall be separately trapped by a water-seal trap, except as otherwise permitted by this code. The vertical distance from the fixture outlet to the trap weir shall not exceed 24 inches (610 mm) and the horizontal distance shall not exceed 30 inches (610 mm) measured from the centerline of the fixture outlet to the centerline of the inlet of the trap. The height of a clothes washer standpipe above a trap shall conform to Section 802.4. A fixture shall not be double trapped.*

Exceptions:

1. *This section shall not apply to fixtures with integral traps.*
2. *A combination plumbing fixture is permitted to be installed on one trap provided that one compartment is not more than 6 inches (152 mm) deeper than the other compartment and the waste outlets are not more than 30 inches (762 mm) apart.*
3. *A grease trap intended to serve as a fixture trap in accordance with the manufacturer's installation instructions shall be permitted to serve as the trap for a single fixture or a combination sink of not more than three compartments where the vertical distance from the fixture outlet to the inlet of the interceptor does not exceed 30 inches (762 mm), and the developed length of the waste pipe from the most upstream fixture outlet to the inlet of the interceptor does not exceed 60 inches (1524 mm).*

Discussion and Commentary: The configuration of a trap interferes with the flow of drainage; however, the interference is minimal because of the construction of the trap and the relatively high inlet velocity of the waste flow. Double trapping of a fixture is prohibited because it will cause air to be trapped between two trap seals, and the *air-bound* drain will impede the flow. The maximum vertical distance of 24 inches (610 mm) from a fixture outlet to the trap weir is required for two reasons. The primary purpose is to reduce odor and the growth of bacteria; therefore, it is desirable to locate the trap as close as possible to the fixture. The vertical distance is also limited to control the velocity of the drainage flow. A maximum horizontal distance of 30 inches must also be adhered to.

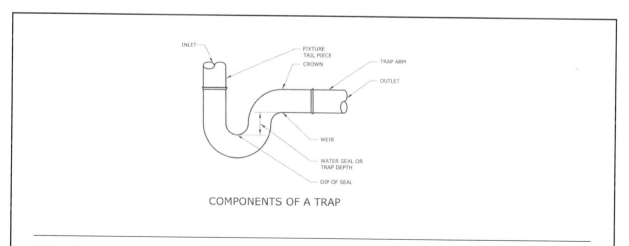

COMPONENTS OF A TRAP

A trap keeps sewer gases from emanating out of the drainage system. The water seal prevents sewer gases and aerosol-borne bacteria from entering the building space. Sewer gases often contain methane gas and could cause an explosion when exposed to an ignition source.

Code Text: *Fixture traps shall be self-scouring. Fixture traps shall not have interior partitions, except where such traps are integral with the fixture or where such traps are constructed of an approved material that is resistant to corrosion and degradation. Slip joints shall be made with an approved elastomeric gasket and shall be installed only on the trap inlet, trap outlet and within the trap seal.*

Discussion and Commentary: A trap must have a pattern allowing unobstructed flow to the drain. Interior partitions are allowable where constructed of a material resistant to corrosion and degradation. A common use would be tight installation areas like pedestal lavatories.

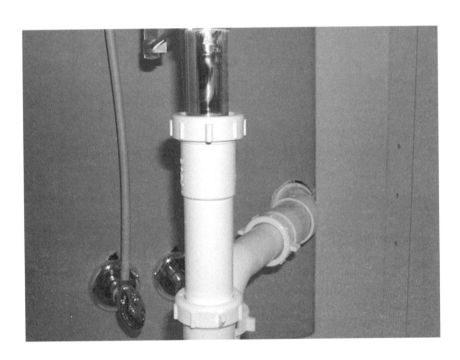

Most traps are premanufactured; however, a trap can be field fabricated with pipe and fittings.

Code Text: *The following types of traps are prohibited:*
1. *Traps that depend on moving parts to maintain the seal.*
2. *Bell traps.*
3. *Crown-vented traps.*
4. *Traps not integral with a fixture and that depend on interior partitions for the seal, except those traps constructed of an approved material that is resistant to corrosion and degradation.*
5. *"S" traps.*
6. *Drum traps.*
   Exception: Drum traps used as solids interceptors and drum traps serving chemical waste systems shall not be prohibited.

Discussion and Commentary: A trap is intended to be a simple U-shaped piping arrangement that offers minimal resistance to flow. Prohibited traps do not incorporate this simple design concept.

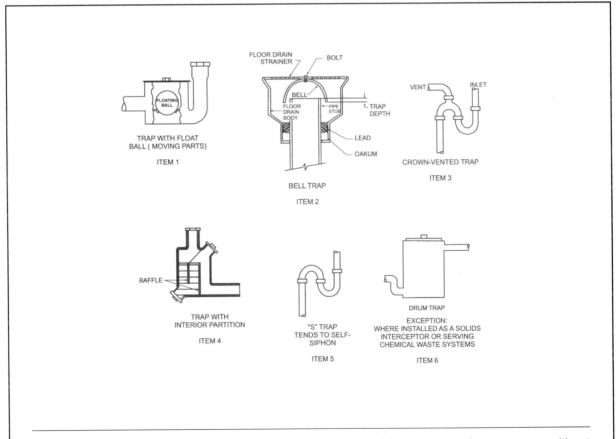

TRAP: A fitting or device that provides a liquid seal to prevent the emission of sewer gases without materially affecting the flow of sewage or wastewater through the trap.

Code Text: *Fixture trap size shall be sufficient to drain the fixture rapidly and not less than the size indicated in Table 709.1. A trap shall not be larger than the drainage pipe into which the trap discharges.*

Discussion and Commentary: The minimum fixture trap sizes are listed in Table 709.1. The minimum trap sizes for fixtures not listed in Table 709.1 must be the size of the fixture outlet, but in no case less than $1^{1}/_{4}$ inches (32 mm). The code does not prescribe a maximum trap size; however, if a trap is oversized, it will not scour (cleanse) itself and will therefore be prone to clogging.

TABLE 709.1
DRAINAGE FIXTURE UNITS FOR FIXTURES AND GROUPS

FIXTURE TYPE	DRAINAGE FIXTURE UNIT VALUE AS LOAD FACTORS	MINIMUM SIZE OF TRAP (inches)
Automatic clothes washers, commercial[a,g]	3	2
Automatic clothes washers, residential[g]	2	2
Bathroom group as defined in Section 202 (1.6 gpf water closet)[f]	5	—
Bathroom group as defined in Section 202 (water closet flushing greater than 1.6 gpf)[f]	6	—
Bathtub[b] (with or without overhead shower or whirlpool attachments)	2	$1^{1}/_{2}$
Bidet	1	$1^{1}/_{4}$
Combination sink and tray	2	$1^{1}/_{2}$

A trap that is larger than the drainage pipe into which it discharges is also subject to frequent clogging because the reduced size outlet pipe will not allow a waste flow velocity that is adequate to scour and clean the trap.

Topic: Building Traps

Reference: IPC 1002.6

Category: Traps, Interceptors and Separators

Subject: Trap Requirements

Code Text: *Building (house) traps shall be prohibited, except where local conditions necessitate such traps. Building traps shall be provided with a cleanout and a relief vent or fresh air intake on the inlet side of the trap. The size of the relief vent or fresh air intake shall not be less than one-half the diameter of the drain to which the relief vent or air intake connects. Such relief vent or fresh air intake shall be carried above grade and shall be terminated in a screened outlet located outside the building.*

Discussion and Commentary: A building trap is not allowed except where needed because of local conditions. Such traps are an obstruction to flow and can cause a complete stoppage in the drainage system. In rare cases, building traps may be necessary where the public sewer exerts positive backpressure on the connected building sewers. This backpressure can cause the building vent terminations to emit strong sewer gases, thereby creating a serious odor problem.

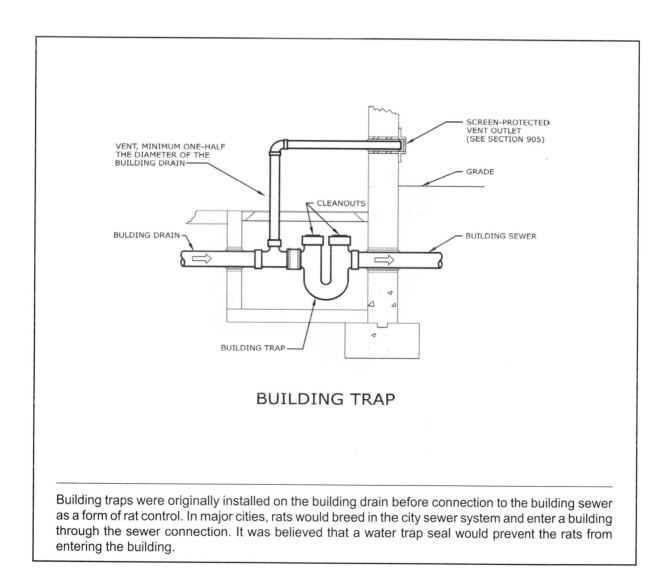

BUILDING TRAP

Building traps were originally installed on the building drain before connection to the building sewer as a form of rat control. In major cities, rats would breed in the city sewer system and enter a building through the sewer connection. It was believed that a water trap seal would prevent the rats from entering the building.

Code Text: *Traps shall be set level with respect to the trap seal and, where necessary, shall be protected from freezing.*

Discussion and Commentary: Traps must be set level to maintain the trap seal depth and reduce the possibility of self-siphoning. Losing the trap seal exposes building occupants to hazards associated with the contents of the sewer drainage system.

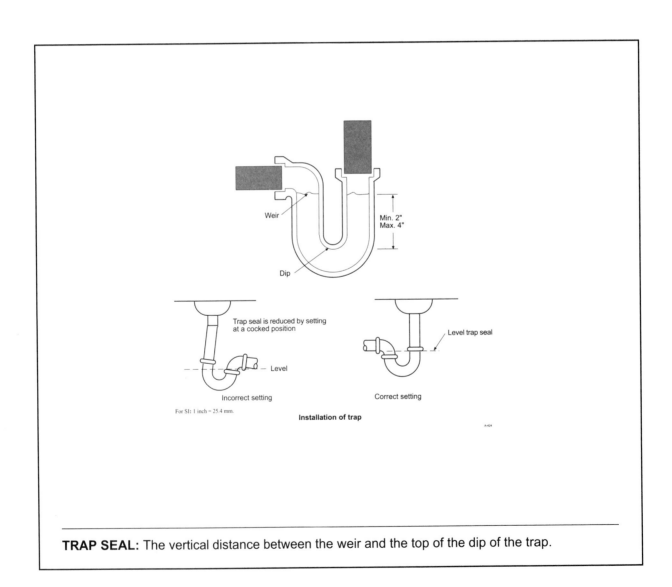

Weir

Min. 2"
Max. 4"

Dip

Trap seal is reduced by setting
at a cocked position

Level

Incorrect setting

For SI: 1 inch = 25.4 mm.

Level trap seal

Correct setting

Installation of trap

TRAP SEAL: The vertical distance between the weir and the top of the dip of the trap.

Code Text: *The size, type and location of each interceptor and of each separator shall be designed and installed in accordance with the manufacturer's instructions and the requirements of this section based on the anticipated conditions of use. Wastes that do not require treatment or separation shall not be discharged into any interceptor or separator.*

Discussion and Commentary: Each interceptor and separator must be designed for the specific installation because the installation depends on the materials being separated. When devices of this specialized nature are involved, no material other than that which requires treatment or separation should be allowed to discharge into the device, unless specifically recommended by the device manufacturer or the design engineer as performing a necessary function. This requires an analysis of the intended use and a determination of the peak load condition.

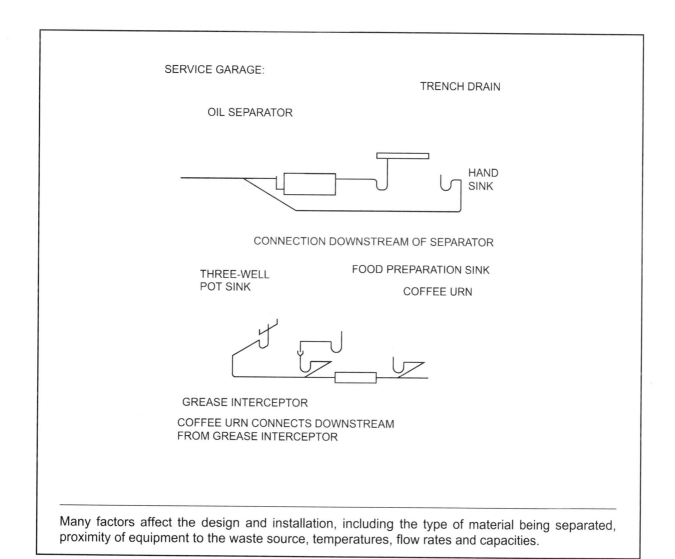

Many factors affect the design and installation, including the type of material being separated, proximity of equipment to the waste source, temperatures, flow rates and capacities.

Code Text: *A grease interceptor or automatic grease removal device shall be required to receive the drainage from fixtures and equipment with grease-laden waste located in food preparation areas, such as in restaurants, hotel kitchens, hospitals, school kitchens, bars, factory cafeterias and clubs. Fixtures and equipment shall include pot sinks, pre-rinse sinks; soup kettles or similar devices; wok stations; floor drains or sinks into which kettles are drained; automatic hood wash units and dishwashers without pre-rinse sinks. Grease interceptors and automatic grease removal devices shall receive waste only from fixtures and equipment that allow fats, oils or grease to be discharged.*

Discussion and Commentary: Interceptors are necessary to remove substances that are detrimental to the drainage system, sewage treatment plants or private sewage disposal systems. Many questions arise as to which fixtures in various occupancies such as restaurants should be connected to the interceptor. The code provides guidance by listing most of such fixtures in this section. Judgment is still required in questionable cases because the code allows only fixtures that allow the discharge of fats to be connected to the interceptor.

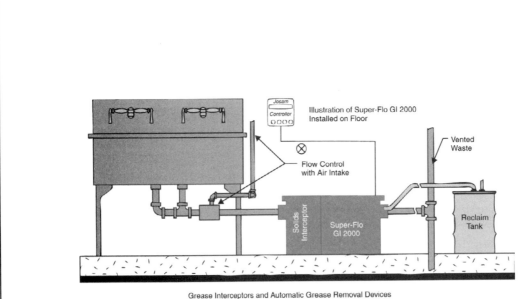

Grease Interceptors and Automatic Grease Removal Devices

Occupancies that commonly require the removal of grease, fat or other solids such as sand include: motor vehicle repair shops, restaurants or cafeterias, hotels, hospitals, animal slaughtering facilities, fowl, fish or meat-packaging plants, and commercial laundries.

Topic: Food Waste Grinders	**Category:** Traps, Interceptors and Separators
Reference: IPC 1003.3.2	**Subject:** Interceptors and Separators

Code Text: *Where food waste grinders connect to grease interceptors, a solids interceptor shall separate the discharge before connecting to the grease interceptor. Solids interceptors and grease interceptors shall be sized and rated for the discharge of the food waste grinder. Emulsifiers, chemicals, enzymes and bacteria shall not discharge into the food waste grinder.*

Discussion and Commentary: A food waste grinder's discharge is permitted to pass through a grease interceptor. It has become increasingly popular for food waste grinders to discharge through grease interceptors to remove grease from food. The interceptor captures grease that would otherwise discharge directly into the drainage system and thus adversely affect the drainage system and waste treatment facilities. Solids interceptors and grease interceptors must be sized and rated for the discharge of the food waste grinder for proper operation of the unit.

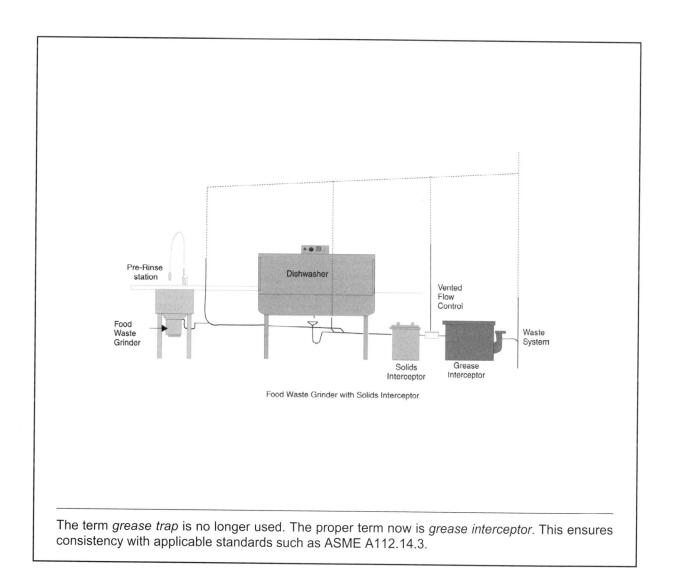

Food Waste Grinder with Solids Interceptor

The term *grease trap* is no longer used. The proper term now is *grease interceptor*. This ensures consistency with applicable standards such as ASME A112.14.3.

Code Text: *Grease interceptors or automatic grease removal devices shall conform to PDI G101, ASME A112.14.3 or ASME A112.14.4 and shall be installed in accordance with the manufacturer's instructions.*

Discussion and Commentary: The referenced standards PDI G101 and ASME A112.14.3 describe a procedure for testing and rating a grease interceptor. The standards do not specify any construction requirements. After testing to the standard, a grease interceptor is certified for its flow rate and grease retention capacity. ASME A112.14.4 is the standard for automatic grease removal devices.

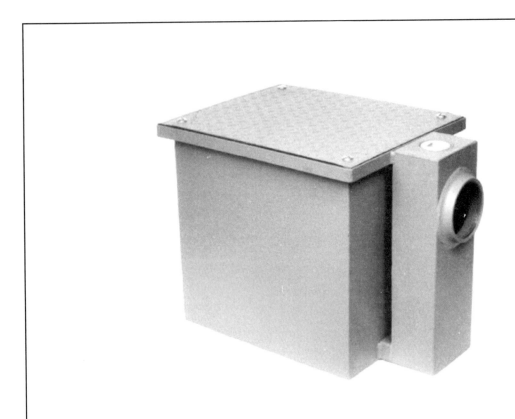

GREASE REMOVAL DEVICE, AUTOMATIC (GRD): A plumbing appurtenance that is installed in the sanitary drainage system to intercept free-floating fats, oils and grease from wastewater discharge. Such a device operates on a time- or event- controlled basis and has the ability to remove free-floating fats, oils and grease automatically without intervention from the user except for maintenance.

Topic: Grease Interceptor Capacity

Category: Traps, Interceptors and Separators

Reference: IPC 1003.3.4.1

Subject: Interceptors and Separators

Code Text: *Grease interceptors shall have the grease retention capacity indicated in Table 1003.3.4.1 for the flow-through rates indicated.*

Discussion and Commentary: Grease interceptor retention capacity must be based on the flow-through rating of the grease interceptor and the discharge rate of the drainage pipe served. The waste flow capacity of a grease interceptor determines the quantity of grease that can be separated from the waste, which, in turn, dictates the required capacity of the grease interceptor to hold the collected grease.

TABLE 1003.3.4.1
CAPACITY OF GREASE INTERCEPTORS [a]

TOTAL FLOW-THROUGH RATING (gpm)	GREASE RETENTION CAPACITY (pounds)
4	8
6	12
7	14
9	18
10	20
12	24
14	28
15	30
18	36
20	40
25	50
35	70
50	100
75	150
100	200

For SI: 1 gallon per minute = 3.785 L/m, 1 pound = 0.454 kg.

a. For total flow-through ratings greater than 100 (gpm) , double the flow-through rating to determine the grease retention capacity (pounds).

The maintenance frequency for grease interceptor is directly proportional to the retention capacity of the device.

Topic: Rate of Flow Controls

Category: Traps, Interceptors and Separators

Reference: IPC 1003.3.4.2

Subject: Interceptors and Separators

Code Text: *Grease interceptors shall be equipped with devices to control the rate of water flow so that the water flow does not exceed the rated flow. The flow-control device shall be vented and terminate not less than 6 inches (152 mm) above the flood rim level or be installed in accordance with the manufacturer's instructions.*

Discussion and Commentary: To prevent an excessive flow rate through the grease interceptor, it must be either large enough to handle the flow, or a flow control device must be installed upstream of the grease interceptor and in accordance with the manufacturer's instructions. The flow control device acts as a restrictor to control the flow into the grease interceptor. Such devices are typically a fitting with a fixed orifice and an air intake or vent and are usually provided and sized by the manufacturer of the grease interceptor.

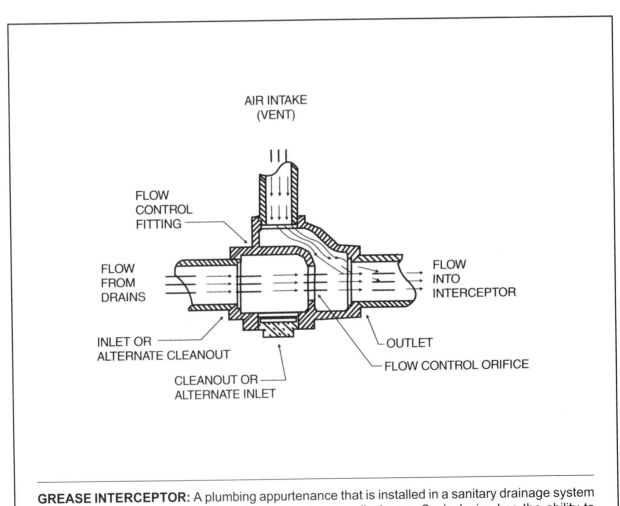

GREASE INTERCEPTOR: A plumbing appurtenance that is installed in a sanitary drainage system to intercept oily and greasy wastes from a wastewater discharge. Such device has the ability to intercept free-floating fats and oils.

Topic: Automatic Grease Removal Devices

Category: Traps, Interceptors and Separators

Reference: IPC 1003.3.5

Subject: Interceptors and Separators

Code Text: *Where automatic grease removal devices are installed, such devices shall be located downstream of each fixture or multiple fixtures in accordance with the manufacturer's instructions. The automatic grease removal device shall be sized to pretreat the measured or calculated flows for all connected fixtures or equipment. Ready access shall be provided for inspection and maintenance.*

Discussion and Commentary: This code text addresses automatic grease removal devices, their sizing and access requirements for inspection and maintenance. Automatic grease removal systems are an acceptable technology when installed properly. The ASME Standard A112.14.4 referenced in Section 1003.3.4 is related to automatic grease removal devices. Ready access must be provided to such devices.

GREASE REMOVAL DEVICE, AUTOMATIC (GRD). A plumbing appurtenance that is installed in the sanitary drainage system to intercept freefloating fats, oils and grease from wastewater discharge. Such a device operates on a time-or event-controlled basis and has the ability to remove freefloating fats, oils and grease automatically without intervention from the user except for maintenance.

READY ACCESS: That which enables a fixture, appliance or equipment to be directly reached without requiring the removal or movement of any panel, door or similar obstruction and without the use of a portable ladder, step stool or similar device.

Code Text: *At repair garages, car-washing facilities, at factories where oily and flammable liquid wastes are produced and in hydraulic elevator pits, separators shall be installed into which all oil-bearing, grease-bearing or flammable wastes shall be discharged before emptying into the building drainage system or other point of disposal.*

Exception: An oil separator is not required in hydraulic elevator pits where an approved alarm system is installed.

Discussion and Commentary: A separator is required in areas used for motor vehicle repairs and car-washing facilities with engine or undercarriage cleaning capability and at factories where oily and flammable liquid wastes are produced. In such facilities, floor drains are the primary receptors of wastes requiring treatment. It is not the intent of this section to require oil separators in a public or private garage. A garage used only as a parking facility is allowed to have floor drains connected to the drainage system without separators.

Oil Separator

(Illustration Courtesy of Rockford Sanitary Systems)

An approved alarm system that provides notification of an oil leak in a hydraulic elevator pit is considered equivalent to installing an oil separator because it would result in the immediate attention to and prevention of an extensive oil spill.

Topic: Design Requirements

Category: Traps, Interceptors and Separators

Reference: IPC 1003.4.2.1

Subject: Oil Separators

Code Text: *Oil separators shall have a depth of not less than 2 feet (610 mm) below the invert of the discharge drain. The outlet opening of the separator shall have not less than an 18-inch (457 mm) water seal.*

Discussion and Commentary: The sizing specified in this section relates to an open-tank-type separator. The minimum depth is necessary to provide retention capacity for sludge and solids and to obtain efficient retention of oil or other volatile liquid wastes by minimizing turbulent flow through the separator. The outlet must be designed to provide at least 18 inches (457 mm) of liquid seal to provide for the storage of oil to at least that depth.

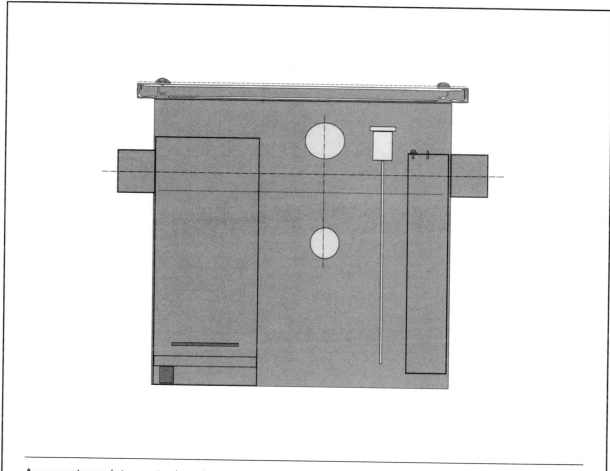

A separator or interceptor is a device designed and installed to separate and retain for removal, by automatic or manual means, deleterious, hazardous or undesirable matter from normal wastes, while permitting normal sewage or wastes to discharge into the drainage system by gravity.

Code Text: *Where automobiles are serviced, greased, repaired or washed or where gasoline is dispensed, oil separators shall have a minimum capacity of 6 cubic feet (0.168m3) for the first 100 square feet (9.3m2) of area to be drained, plus 1 cubic foot (0.28 m3) for each additional 100 square feet (9.3 m2) of area to be drained into the separator. Parking garages in which servicing, repairing or washing is not conducted, and in which gasoline is not dispensed, shall not require a separator. Areas of commercial garages utilized only for storage of automobiles are not required to be drained through a separator.*

Discussion and Commentary: In addition to the general requirements of Section 1003.4.2.1, this section specifically applies to motor vehicle garages or service stations where lubrication, oil changing, fuel dispensing, repair work, or hand or mechanical washing takes place. The intent of this section is to require separators in locations where flammable or combustible liquids are to be discharged into the drainage system, typically in occupancies provided with floor or trench drains.

The requirement for a separator does not, in itself, create the requirement for floor or trench drains. Where a drainage system is provided in these occupancies, provisions must be made to prevent flammable and combustible liquids from entering the drainage system, sewers and waste treatment facilities.

Code Text: *Laundry facilities not installed within an individual dwelling unit or intended for individual family use shall be equipped with an interceptor with a wire basket or similar device, removable for cleaning, that prevents passage into the drainage system of solids 0.5 inch (12.7 mm) or larger in size, string, rags, buttons or other materials detrimental to the public sewage system.*

Discussion and Commentary: Commercial laundries and similar establishments must be equipped with an interceptor that is capable of preventing string, lint and other solids from entering the sewage system. The filter, screen or basket must allow for cleaning and removing intercepted solids.

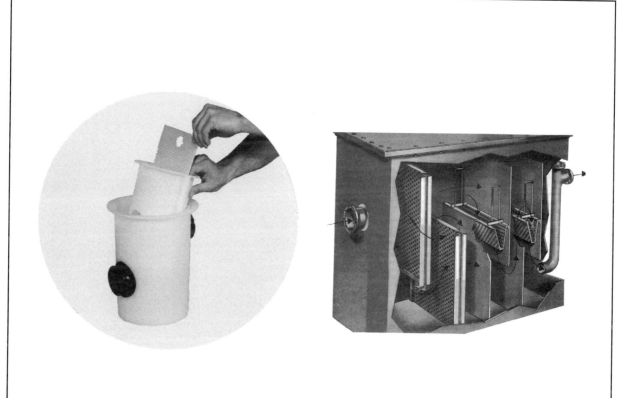

DWELLING: Any building that contains one or two dwelling units used, intended or designed to be built, used, rented, leased, let or hired out to be occupied, or that are occupied for living purposes (IRC Section R202).

Code Text: *Interceptors and separators shall be designed so as not to become air bound where tight covers are utilized. Each interceptor or separator shall be vented where subject to a loss of trap seal.*

Discussion and Commentary: Most interceptors and separators require venting of the containment tank. Typically, a vent is required to allow the escape or admittance of air to compensate for the variable fluid level in the tank. Oil and gasoline separators require independent vents to help control and dissipate vapor buildup that may occur in the holding tank.

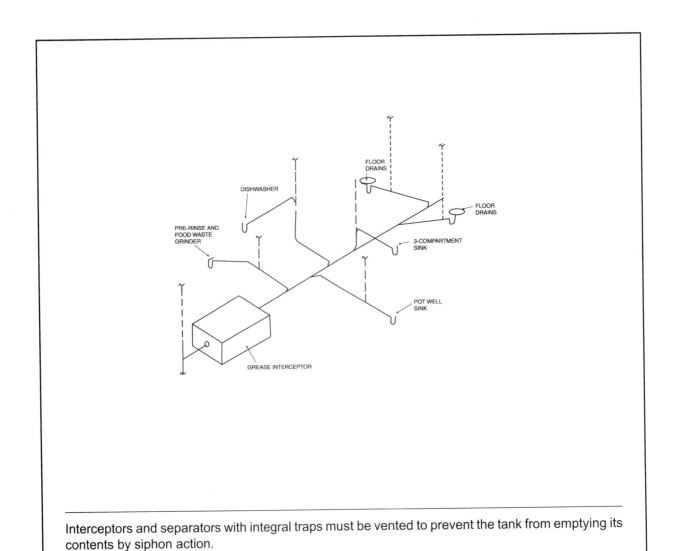

Interceptors and separators with integral traps must be vented to prevent the tank from emptying its contents by siphon action.

Quiz

Study Session 14
IPC Chapter 10

1. Which of the following statements is *true* regarding the trapping of a three-compartment sink in a commercial kitchen?

   a. The vertical distance from the fixture outlet to the trap weir cannot exceed 6 inches.

   b. The sink is permitted to be installed on one trap, using a center outlet continuous waste connection.

   c. The sink is permitted to be installed on one trap, using an end outlet continuous waste to connect the three compartments.

   d. The maximum horizontal distance from any one outlet to the trap inlet cannot exceed 12 inches.

   Reference ___B___ SEC 1002

2. Which of the following traps is approved on a dental cuspidor, where it is designed to capture solids such as dental gold, silver or porcelain?

   a. bell trap   b. S trap

   c. drum trap   d. crown-vented trap

   Reference ___C___ 102.3

3. Where are exposed traps prohibited?

    a. trauma centers   b. handicap accessible restrooms

    c. food service establishments   (d.) mental health facilities

Reference _____D_____

4. All of the following are requirements for trap design and installation, *except* _____.

    a. the trap seal must be level

    b. elastomeric slip joints are only allowed on the trap inlet

    c. the trap seal must have a two to four inch liquid depth

    d. where a trap seal is subject to loss by evaporation, a trap seal primer valve is required

Reference _____   SEC. 102.2   102.4

5. Which trap design is prohibited in all applications?

    (a.) a trap designed with a floating ball-check

    b. interceptor type drum trap

    c. a trap integral to the fixture

    d. a deep seal trap

Reference _____

6. The minimum trap size for a tub with an overhead shower is _____ inches.

    a. $1^{1}/_{4}$ (b.) $1^{1}/_{2}$

    c. 2   d. 3

Reference _____

7. In some cases under special local conditions, a(n) _____ trap is permitted.

    a. bell   b. S

    c. crown-vented   (d.) building

Reference _____   102.6.

8. The installation of a grease interceptor that is manufactured to serve as the trap for the fixture drain is limited to _____ .

    a. serving as a fixture trap for no more than two fixture drains

    b. a maximum of 60 inches in developed length from the fixture outlet to the inlet of the interceptor

    c. a maximum of 30 inches horizontally from the centerline of the fixture outlet to the inlet of the interceptor

    d. a maximum of 24 inches vertically from the fixture outlet to the inlet of the interceptor

Reference _____ 102.1

9. Where the trap seal is dependent on an interior partition, only approved traps constructed of corrosion resistant material or _____ are permitted.

    a. traps integral to the fixture

    b. readily accessible traps

    c. traps with PO plugs

    d. self-scouring traps with slip joint connections

Reference _____

10. Where the manufacturer's design permits, a grease interceptor is allowed to serve as a fixture trap for how many fixtures?

    a. one   b. two

    c. three   d. four

Reference ___A___ 102.1

11. Substances that are detrimental to the sewer system are prevented from being discharged into the drainage system by _____ .

    a. filters   b. strainers

    c. traps or backwater valves   d. interceptors or separators

Reference _____

12. Which of the following fixtures in a food preparation facility is not allowed to discharge to a grease interceptor?

    a. dishwasher

    b. steam table

    c. scullery sink

    d. kitchen area drains

    Reference _____ 103.2

13. A food waste grinder is prohibited from discharging to a grease interceptor without _____ .

    a. being treated with emulsifiers

    b. cold water dilution and enzyme treatment

    c. first discharging through a solids interceptor

    d. first discharging through an oil separator

    Reference _____ 103.3-2

14. A grease interceptor is not required for the kitchen of a _____ .

    a. factory

    b. hospital

    c. school

    d. dwelling unit

    Reference _____

15. An auto repair garage has an oil separator receiving drainage from a repair bay that has an area of 100 feet x 27 feet. What is the minimum capacity required for the oil separator to serve the garage work area?

    a. 6 cu ft

    b. 27 cu ft

    c. 32 cu ft

    d. 40 cu ft

    Reference ___C___

16. The water seal for a sand interceptor shall be not less than _____ inches.

    a. 6

    b. 8

    c. 12

    d. 18

    Reference 103.5

17. The minimum height of the vent termination for a flow control device on a grease interceptor is six inches above the _____ .

    a.   roof of the building.

    b.   flood rim of the interceptor

    c.   flood rim of the connected fixture

    d.   drainage connection to the interceptor

Reference _____

18. Automatic grease removal devices are required to be sized according to the calculated _____ .

    a.   flows during peak usage

    b.   flows of all connected fixtures

    c.   average flows of the fixtures

    d.   flow rate from the highest volume fixture

Reference _____

19. Which of the following is required to prevent the loss of the trap seal on an interceptor or separator?

    a.   floating cover

    b.   air tight cover

    c.   venting of the flow control device

    d.   venting of the drain from the interceptor

Reference _____ 103.9

20. The vertical distance from the fixture outlet to the trap weir shall not exceed _____ inches.

    a.   18   b.   24

    c.   30   d.   36

Reference _____

21. Each fixture trap shall have a liquid seal of not less than _____ inches and not more than _____ inches.

    a. 2, 6   b. 1, 4

    c. 2, 4   d. 2, 5

Reference _____

22. Where an acid-resisting trap is installed underground, such trap shall be embedded in concrete extending _____ inches beyond the bottom and sides of the trap.

    a. 2   b. 3

    c. 4   d. 6   *102.9*

Reference _____

23. The capacity of a grease trap with a total flow-through rating of 10 gpm is _____ pounds.

    a. 14   b. 18

    c. 20   d. 24

Reference *10-18*

24. Oil separators shall have a depth of not less than _____ inches below the invert of the discharging drain.

    a. 24   b. 36

    c. 30   d. 42

Reference _____

25. The outlet opening of an oil separator shall have a water seal of not less than _____ inches.

    a. 6   b. 12

    c. 18   d. 24

Reference *103.4*

2006 IPC Chapters 11 and 12
Storm Drainage and Special Piping and Storage Systems

OBJECTIVE: To develop an understanding of the code provisions applicable to the removal of water associated with storm and rainfall and of the precautions against the structural failure of a flat roof. To develop an understanding of the code provisions that apply to systems for nonflammable medical gas and nonmedical oxygen systems.

REFERENCE: Chapters 11 and 12, 2006 *International Plumbing Code*

KEY POINTS:
- Are reductions in the drainage pipe size allowed for storm drainage systems?
- When can storm water drain into a sanitary sewer?
- What requirements apply to the fittings of the system?
- What pipe material is permitted for use as subsoil storm drain pipe, and what standard applies?
- Is vitrified clay pipe permitted for use as subsoil drainpipe?
- What controls the trap size?
- What type of cleanout is required, and where is it to be located?
- When a combined sewer is utilized, where is the storm drain connected and with what type of fitting?
- Are floor drains permitted to be connected to a storm drain?
- How high above the surface are roof drain strainers required to extend?
- What rainfall criteria are used for the sizing of a storm drainage system?
- Which table is used for sizing vertical storm drainage leaders, and on what is the size based?
- How are vertical walls taken into consideration when sizing roof drains?
- Which code controls the location of parapet wall scuppers?
- How are semicircular gutters sized?
- What condition triggers the need of a secondary drainage system?
- Does the code allow a common point of discharge?
- Is the flow through the primary system considered when sizing the secondary system?
- How many square feet are equal to one fixture unit when the area is greater than 4,000 square feet?

KEY POINTS:
(Cont'd)

- Control devices are required for what purpose?
- What type of protection is required for control devices?
- How many roof drains are required for a roof area not over 10,000 square feet?
- When is an accessible backwater valve required?
- When is a sump required?
- What is the minimum diameter of the sump pit?
- What provisions apply to electrical service outlets?
- Which NFPA standard applies to nonmedical gases?
- Which NFPA standards apply to the design and installation of nonmedical oxygen systems?

Code Text: *The size of a drainage pipe shall not be reduced in the direction of flow.*

Discussion and Commentary: One of the fundamental requirements of any form of drainage system is that the piping is not permitted to be reduced in size in the direction of drainage flow. This is also the case with storm drainage systems. A size reduction creates an obstruction to the flow of discharge, possibly resulting in a backup, an interruption of service in the drainage system or a stoppage occurring in the piping.

Storm drainage system is a drainage system that carries rainwater, surface water, subsurface water and similar liquid wastes.

Code Text: *Roofs shall be designed for the maximum possible depth of water that will pond thereon as determined by the relative levels of roof deck and overflow weirs, scuppers, edges or serviceable drains in combination with the deflected structural elements. In determining the maximum possible depth of water, all primary roof drainage means shall be assumed to be blocked.*

Discussion and Commentary: The roof structure must be capable of supporting the maximum ponding of water that will occur when the primary roof drainage means are blocked. In Section 1107, the code requires the installation of a secondary (emergency) roof drainage system to limit the amount of rainwater that could be retained on top of the roof. This also limits the potential increased load to be structurally supported by the roof.

DRAINAGE SYSTEM	FLOW RATE (gpm)									
	Depth of water above drain inlet (hydraulic head) (inches)									
	1	2	2.5	3	3.5	4	4.5	5	7	8
4-inch-diameter drain	80	170	180							
6-inch-diameter drain	100	190	270	380	540					
8-inch-diameter drain	125	230	340	560	850	1,100	1,170			
6-inch-wide, open-top scupper	18	50	*	90	*	140	*	194	321	393
24-inch-wide, open-top scupper	72	200	*	360	*	560	*	776	1,284	1,572
6-inch-wide, 4-inch-high, closed-top scupper	18	50	*	90	*	140	*	177	231	253
24-inch-wide, 4-inch-high, closed-top scupper	72	200	*	360	*	560	*	708	924	1,012
6-inch-wide, 6-inch-high, closed-top scupper	18	50	*	90	*	140	*	194	303	343
24-inch-wide, 6-inch-high, closed-top scupper	72	200	*	360	*	560	*	776	1,212	1,372

For SI: 1 inch = 25.4 mm, 1 gallon per minute = 3.785 L/m.

Source: Factory Mutual Engineering Corp. Loss Prevention Data 1-54.

There have been documented cases of roof structural collapse that are due to roof drain blockage causing excessive rain water ponding.

Topic: Subsoil Drain Pipe

Category: Storm Drainage

Reference: IPC 1102.5

Subject: Materials

Code Text: *Subsoil drains shall be open-jointed, horizontally split or perforated pipe conforming to one of the standards listed in Table 1102.5.*

Discussion and Commentary: The referenced table specifies acceptable piping materials for subsoil drainage systems. Each material must comply with its own specific standard as shown in the table. All of the criteria and limitations of the standards are applicable unless otherwise specified in the code.

TABLE 1102.5
SUBSOIL DRAIN PIPE

MATERIAL	STANDARD
Asbestos-cement pipe	ASTM C 508
Cast-iron pipe	ASTM A 74; ASTM A 888; CISPI 301
Polyethylene (PE) plastic pipe	ASTM F 405; CSA B182.1; CSA B182.6; CSA B182.8
Polyvinyl chloride (PVC) Plastic pipe (type sewer pipe, PS25, PS50 or PS100)	ASTM D 2729; ASTM F 891; CSA B182.2; CSA B182.4
Stainless steel drainage systems, Type 316L	ASME A112.3.1
Vitrified clay pipe	ASTM C 4; ASTM C 700

SUBSOIL DRAIN: A drain that collects subsurface water or seepage water and conveys such water to a place of disposal.

Code Text: *Pipe fittings shall be approved for installation with the piping material installed, and shall conform to the respective pipe standards or one of the standards listed in Table 1102.7. The fittings shall not have ledges, shoulders or reductions capable of retarding or obstructing flow in the piping. Threaded drainage pipe fittings shall be of the recessed drainage type.*

Discussion and Commentary: Each fitting is designed to be installed in a particular system with a given material or combination of materials. Drainage pattern fittings must be installed in drainage systems. Vent fittings are limited to the venting system. There are also fittings available that could be used in any type of plumbing system. Fittings must be the same material as, or compatible with, the pipe. This prevents any chemical or corrosive action between dissimilar materials.

TABLE 1102.7
PIPE FITTINGS

MATERIAL	STANDARD
Acrylonitrile butadiene styrene (ABS) plastic	ASTM D 2661; ASTM D 3311; CSA B181.1
Cast-iron	ASME B16.4; ASME B16.12; ASTM A 888; CISPI 301; ASTM A 74
Coextruded composite ABS sewer and drain DR-PS in PS35, PS50, PS100, PS140, PS200	ASTM D 2751
Coextruded composite ABS DWV Schedule 40 IPS pipe (solid or cellular core)	ASTM D 2661; ASTM D 3311; ASTM F 628
Coextruded composite PVC DWV Schedule 40 IPS-DR, PS140, PS200 (solid or cellular core)	ASTM D 2665; ASTM D 3311; ASTM F 891
Coextruded composite PVC sewer and drain DR-PS in PS35, PS50, PS100, PS140, PS200	ASTM D 3034
Copper or copper alloy	ASME B16.15; ASME B16.18; ASME B16.22; ASME B16.23; ASME B16.26; ASME B16.29
Gray iron and ductile iron	AWWA C110
Malleable iron	ASME B16.3
Plastic, general	ASTM F 409
Polyvinyl chloride (PVC) plastic	ASTM D 2665; ASTM D 3311; ASTM F 1866
Steel	ASME B16.9; ASME B16.11; ASME B16.28
Stainless steel drainage Systems, Type 316L	ASME A112.3.2

Many pipe standards also include provisions regulating fittings. There are, however, a number of standards that strictly regulate pipe fittings.

Topic: Floor Drains	Category: Storm Drainage
Reference: IPC 1104.3	Subject: Conductors and Connections

Code Text: *Floor drains shall not be connected to a storm drain.*

Discussion and Commentary: To prevent contamination of storm water discharge with chemicals, waste or sewage, the code prohibits the connection of floor drains to a storm drain. A floor drain connected to a storm drainage system invites the discharge of sanitary waste into the storm system, which, considering that storm drainage is discharged to the environment without treatment, would create an environmental hazard.

DRAINAGE SYSTEM. Piping within a public or private premise that conveys sewage, rainwater or other liquid wastes to a point of disposal. A drainage system does not include the mains of a public sewer system or a private or public sewage treatment or disposal plant.

Storm. A drainage system that carries rainwater, surface water, subsurface water and similar liquid wastes.

Areaway drains serving subsurface spaces outside of the building, such as window wells or basement or cellar entrance wells, are not considered to be floor drains when applying this section.

Topic: Strainers

Category: Storm Drainage

Reference: IPC 1105.1

Subject: Roof Drains

Code Text: *Roof drains shall have strainers extending not less than 4 inches (102 mm) above the surface of the roof immediately adjacent to the roof drain. Strainers shall have an available inlet area, above roof level, of not less than one and one-half times the area of the conductor or leader to which the drain is connected.*

Discussion and Commentary: A strainer, sometimes called a *beehive strainer*, helps prevent leaves, debris and roof ballast materials from entering the storm drain. The beehive shape of the strainer helps prevent the entire opening of the strainer from being obstructed when there is an accumulation of debris. This section regulates how far the strainer should extend above the roof. The requirement for the inlet area to be at least one and one-half times the size of the opening (150 percent) provides some assurance that the roof drain will continue to flow at full capacity when the strainer is partially obstructed.

ROOF DRAIN: A drain installed to receive water collecting on the surface of a roof and to discharge such water into a leader or a conductor.

Topic: Flat Decks

Category: Storm Drainage

Reference: IPC 1105.2

Subject: Roof Drains

Code Text: *Roof drain strainers for use on sun decks, parking decks and similar areas that are normally serviced and maintained shall comply with Section 1105.1 or shall be of the flat-surface type, installed level with the deck, with an available inlet area not less than two times the area of the conductor or leader to which the drain is connected.*

Discussion and Commentary: Where a roof deck is used for parking, recreation and other purposes, the roof drain must be flat. It is assumed that any blockage would be noticed by individuals frequenting the roof. Where this flat-surface-type drain is used, the amount of available inlet area must increase to a minimum of two times (200 percent) the area of the attached conductor or leader.

Flat surface type roof drains are designed such that an obstruction to vehicular traffic or tripping hazard to pedestrian traffic does not occur.

Code Text: *Vertical conductors and leaders shall be sized for the maximum projected roof area, in accordance with Table 1106.2.*

Discussion and Commentary: Unlike the flow in horizontal pipe, it is not possible to have full flow in a vertical pipe under gravity flow conditions. The sizing information, including the projected roof area, the diameter of the leader and the rainfall rates, listed in Table 1106.2, is based on the maximum probable capacity of a vertical pipe, which is approximately 29 percent of the pipe's cross-sectional area.

TABLE 1106.2
SIZE OF VERTICAL CONDUCTORS AND LEADERS

DIAMETER OF OF LEADER (inches)[a]	HORIZONTALLY PROJECTED ROOF AREA (square feet)											
	Rainfall rate (inches per hour)											
	1	2	3	4	5	6	7	8	9	10	11	12
2	2,880	1,440	960	720	575	480	410	360	320	290	260	240
3	8,800	4,400	2,930	2,200	1,760	1,470	1,260	1,100	980	880	800	730
4	18,400	9,200	6,130	4,600	3,680	3,070	2,630	2,300	2,045	1,840	1,675	1,530
5	34,600	17,300	11,530	8,650	6,920	5,765	4,945	4,325	3,845	3,460	3,145	2,880
6	54,000	27,000	17,995	13,500	10,800	9,000	7,715	6,750	6,000	5,400	4,910	4,500
8	116,000	58,000	38,660	29,000	23,200	19,315	16,570	14,500	12,890	11,600	10,545	9,600

For SI: 1 inch = 25.4 mm, 1 square foot = 0.0929 m².

a. Sizes indicated are the diameter of circular piping. This table is applicable to piping of other shapes provided the cross-sectional shape fully encloses a circle of the diameter indicated in this table.

Rainfall rates for various cities in the United States are listed in the IPC Appendix B.

Code Text: *The size of the building storm drain, building storm sewer and their horizontal branches having a slope of one-half unit or less vertical in 12 units horizontal (4-percent slope) shall be based on the maximum projected roof area in accordance with Table 1106.3. The minimum slope of horizontal branches shall be one-eighth unit vertical in 12 units horizontal (1-percent slope) unless otherwise approved.*

Discussion and Commentary: Unlike a sanitary drainage system, a horizontal storm drain or sewer is sized for a full-flow condition. The pipe, may be filled to capacity under a worst-case condition, is still sized for gravity flow. Under normal operation, the storm sewer is not full and functions more like a sanitary sewer.

TABLE 1106.3
SIZE OF HORIZONTAL STORM DRAINGE PIPING

SIZE OF HORIZONTAL PIPING (inches)	HORIZONTALLY PROJECTED ROOF AREA (square feet)					
	Rainfall rate (inches per hour)					
	1	2	3	4	5	6
¹/₈ unit vertical in 12 units horizontal (1-percent slope)						
3	3,288	1,644	1,096	822	657	548
4	7,520	3,760	2,506	1,800	1,504	1,253
5	13,360	6,680	4,453	3,340	2,672	2,227
6	21,400	10,700	7,133	5,350	4,280	3,566
8	46,000	23,000	15,330	11,500	9,200	7,600
10	82,800	41,400	27,600	20,700	16,580	13,800
12	133,200	66,600	44,400	33,300	26,650	22,200
15	218,000	109,000	72,800	59,500	47,600	39,650
¹/₄ unit vertical in 12 units horizontal (2-percent slope)						
3	4,640	2,320	1,546	1,160	928	773
4	10,600	5,300	3,533	2,650	2,120	1,766
5	18,880	9,440	6,293	4,720	3,776	3,146
6	30,200	15,100	10,066	7,550	6,040	5,033
8	65,200	32,600	21,733	16,300	13,040	10,866
10	116,800	58,400	38,950	29,200	23,350	19,450
12	188,000	94,000	62,600	47,000	37,600	31,350
15	336,000	168,000	112,000	84,000	67,250	56,000

When sizing a storm drainage system, the local rainfall rate must be used. Table 1106.3 and maps of the United States indicate the rainfall rates for a storm of 1-hour duration and a 100-year return period.

Code Text: *In sizing roof drains and storm drainage piping, one-half of the area of any vertical wall that diverts rainwater to the roof shall be added to the projected roof area for inclusion in calculating the required size of vertical conductors, leaders and horizontal storm drainage piping.*

Discussion and Commentary: This section includes the requirement for the mandatory inclusion of half of the area of vertical walls, including parapet walls, that are adjacent to and above a roof area, into the calculation for projected horizontal roof area when sizing storm drainage components. This requirement acknowledges that such vertical walls can catch and divert rainwater onto the roof. As such, this additional water must be accounted for in the sizing of the components that will catch and discharge this water.

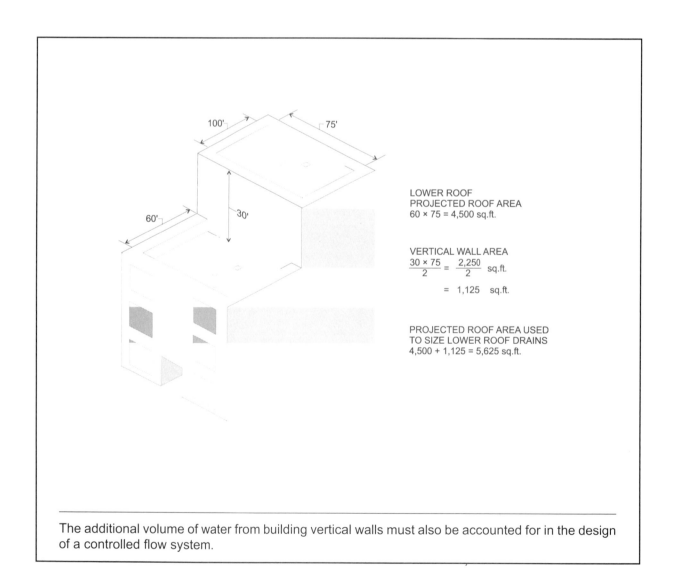

LOWER ROOF
PROJECTED ROOF AREA
60 × 75 = 4,500 sq.ft.

VERTICAL WALL AREA
$$\frac{30 \times 75}{2} = \frac{2,250}{2} \text{ sq.ft.}$$
$$= 1,125 \text{ sq.ft.}$$

PROJECTED ROOF AREA USED
TO SIZE LOWER ROOF DRAINS
4,500 + 1,125 = 5,625 sq.ft.

The additional volume of water from building vertical walls must also be accounted for in the design of a controlled flow system.

Code Text: *The size of semicircular gutters shall be based on the maximum projected roof area in accordance with Table 1106.6.*

Discussion and Commentary: By placing the mechanical equipment in the attic, it is possible to better utilize the building's floor area. However, it is important that such equipment be accessible for service or removal. In addition to a properly sized access opening, a continuous passageway of solid flooring must be provided where needed to reach the appliance. A minimum 30-inch by 30-inch service area is required in front of any appliance access.

TABLE 1106.6
SIZE OF SEMICIRCULAR ROOF GUTTERS

DIAMETER OF GUTTERS (inches)	HORIZONTALLY PROJECTED ROOF AREA (square feet)					
	Rainfall rate (inches per hour)					
	1	2	3	4	5	6
$^1/_{16}$ unit vertical in 12 units horizontal (0.5-percent slope)						
3	680	340	226	170	136	113
4	1,440	720	480	360	288	240
5	2,500	1,250	834	625	500	416
6	3,840	1,920	1,280	960	768	640
7	5,520	2,760	1,840	1,380	1,100	918
8	7,960	3,980	2,655	1,990	1,590	1,325
10	14,400	7,200	4,800	3,600	2,880	2,400
$^1/_8$ unit vertical 12 units horizontal (1-percent slope)						
3	960	480	320	240	192	160
4	2,040	1,020	681	510	408	340
5	3,520	1,760	1,172	880	704	587
6	5,440	2,720	1,815	1,360	1,085	905
7	7,800	3,900	2,600	1,950	1,560	1,300
8	11,200	5,600	3,740	2,800	2,240	1,870
10	20,400	10,200	6,800	5,100	4,080	3,400
$^1/_4$ unit vertical in 12 units horizontal (2-percent slope)						
3	1,360	680	454	340	272	226
4	2,880	1,440	960	720	576	480
5	5,000	2,500	1,668	1,250	1,000	834
6	7,680	3,840	2,560	1,920	1,536	1,280
7	11,040	5,520	3,860	2,760	2,205	1,840
8	15,920	7,960	5,310	3,980	3,180	2,655
10	28,800	14,400	9,600	7,200	5,750	4,800
$^1/_2$ unit vertical in 12 units horizontal (4-percent slope)						
3	1,920	960	640	480	384	320
4	4,080	2,040	1,360	1,020	816	680
5	7,080	3,540	2,360	1,770	1,415	1,180
6	11,080	5,540	3,695	2,770	2,220	1,850
7	15,600	7,800	5,200	3,900	3,120	2,600
8	22,400	11,200	7,460	5,600	4,480	3,730
10	40,000	20,000	13,330	10,000	8,000	6,660

For SI: 1 inch = 25.4 mm, 1 square foot = 0.0929 m².

The size and number of conductors or leader connections to a roof gutter system must be adequate to prevent overflow of the gutters.

Code Text: *Secondary (emergency) roof drain systems shall be sized in accordance with Section 1106 based on the rainfall rate for which the primary system is sized in Tables 1106.2, 1106.3 and 1106.6. Scuppers shall be sized to prevent the depth of ponding water from exceeding that for which the roof was designed as determined by Section 1101.7. Scuppers shall not have an opening dimension of less than 4 inches (102 mm). The flow through the primary system shall not be considered when sizing the secondary roof drain system.*

Discussion and Commentary: The sizing tables contained in Chapter 11, Tables 1106.2, 1106.3 and 1106.6, are based on handling the anticipated rainfall rate (in inches per hour) for a particular horizontal projected roof area. This area is to include at least 50 percent of the area of adjacent vertical walls that can convey rainwater onto the roof. The actual process of sizing the components of a secondary drainage system is identical to sizing the primary system. The same tables and rainfall rates are used. Instead of installing a secondary system of roof drains and pipes, many designers install scuppers to allow rainwater to overflow the roof.

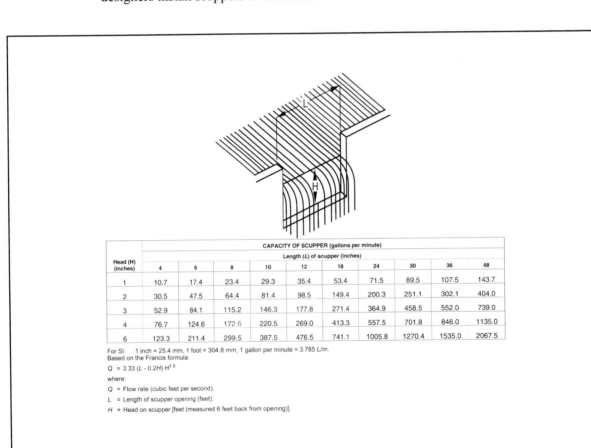

Head (H) (inches)	CAPACITY OF SCUPPER (gallons per minute)									
	Length (L) of scupper (inches)									
	4	6	8	10	12	18	24	30	36	48
1	10.7	17.4	23.4	29.3	35.4	53.4	71.5	89.5	107.5	143.7
2	30.5	47.5	64.4	81.4	98.5	149.4	200.3	251.1	302.1	404.0
3	52.9	84.1	115.2	146.3	177.8	271.4	364.9	458.5	552.0	739.0
4	76.7	124.6	172.6	220.5	269.0	413.3	557.5	701.8	846.0	1135.0
6	123.3	211.4	299.5	387.5	476.5	741.1	1005.8	1270.4	1535.0	2067.5

For SI: 1 inch = 25.4 mm, 1 foot = 304.8 mm, 1 gallon per minute = 3.785 L/m.
Based on the Francis formula:

$Q = 3.33 (L - 0.2H) H^{1.5}$

where:

Q = Flow rate (cubic feet per second).

L = Length of scupper opening (feet).

H = Head on scupper [feet (measured 6 feet back from opening)].

Although the code states that the minimum scupper dimension is 4 inches (102 mm), it would be necessary to increase the length dimension to keep the ponding depth at or below the level for which the roof was designed.

Code Text: *Where there is a continuous or semi-continuous discharge into the building storm drain or building storm sewer, such as from a pump, ejector, air conditioning plant or similar device, each gallon per minute (L/m) of such discharge shall be computed as being equivalent to 96 square feet (9 m2) of roof area, based on a rainfall rate of 1 inch (25.4 mm) per hour.*

Discussion and Commentary: This section regulates how to size piping that receives a continuous or semicontinuous discharge lead. Continuous flow is not converted to the local rainfall rate because of the constant nature of its discharge and is considered in the same manner as the dfu area in combined sewers.

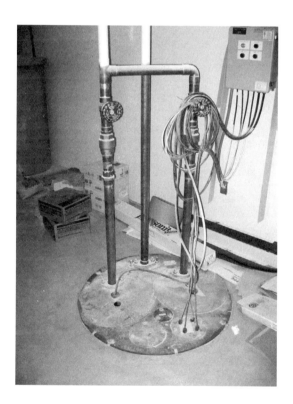

Pump discharge capacity (in gpm) × 96 = Equivalent projected roof area in square feet or metric conversion.
Pump discharge (in lpm) × 2.356 = Equivalent projected roof area in square meters.

Code Text: *Not less than two roof drains shall be installed in roof areas 10,000 square feet (929 m²) or less and not less than four roof drains shall be installed in roofs over 10,000 square feet (929 m²) in area.*

Discussion and Commentary: For controlled flow systems, a system redundancy safety precaution is built into the design by the requirement for a minimum of two drains for roof areas of 10,000 square feet (929 m²) or less. The objective is that if one drain becomes clogged, the other would allow water to drain. For larger buildings in excess of 10,000 square feet (929 m²) of roof area, a minimum of four roof drains is required.

CONTROLLED FLOW DRAINAGE
AREA: 8,000 SQ.FT.
MINIMUM 2 ROOF DRAINS

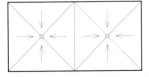

CONTROLLED FLOW DRAINAGE
AREA: 45,000 SQ.FT.
MINIMUM 4 ROOF DRAINS

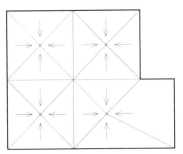

Although no specific requirements exist in this section about how to design the controlled flow system, the potential for partial clogging of part of the system must be evaluated so that the contained water will drain off the roof without any structural or other problems.

Topic: Subsoil Drains	**Category:** Storm Drainage
Reference: IPC 1111.1	**Subject:** Controlled Flow Systems

Code Text: *Subsoil drains shall be open-jointed, horizontally split or perforated pipe conforming to one of the standards listed in Table 1102.5. Such drains shall not be less than 4 inches (102 mm) in diameter. Where the building is subject to backwater, the subsoil drain shall be protected by an accessibly located backwater valve. Subsoil drains shall discharge to a trapped area drain, sump, dry well or approved location above ground. The subsoil sump shall not be required to have either a gas-tight cover or a vent. The sump and pumping system shall comply with Section 1113.1.*

Discussion and Commentary: Table 1102.5 summarizes the applicable standards for subsoil drain pipes. The pipes listed in the table include Asbestos-Cement, Cast-Iron, PE, PVC, Stainless Steel and Vitrified Clay, mostly regulated with the applicable ASTM or CSA Standards. The minimum allowable size for a subsoil drain pipe is 4 inches (102 mm).

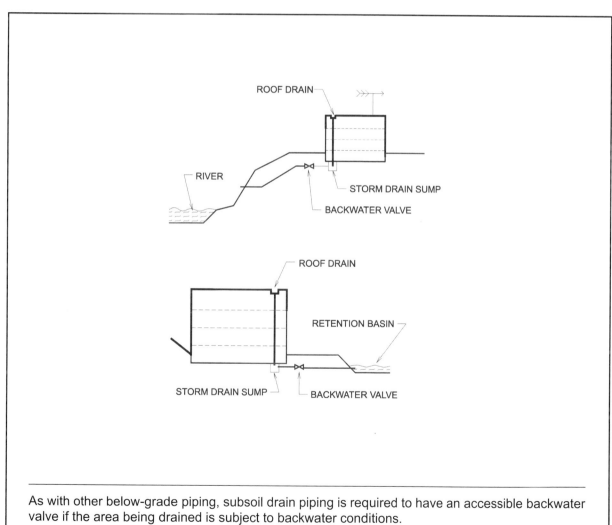

As with other below-grade piping, subsoil drain piping is required to have an accessible backwater valve if the area being drained is subject to backwater conditions.

Code Text: *Nonflammable medical gas systems, inhalation anesthetic systems and vacuum piping systems shall be designed and installed in accordance with NFPA 99C.*

Exceptions:
1. *This section shall not apply to portable systems or cylinder storage.*
2. *Vacuum system exhaust shall comply with the* International Mechanical Code.

Discussion and Commentary: NFPA 99C applies to all nonflammable medical gas, inhalation anesthetic and permanently installed vacuum piping systems, with the exception of portable systems and the storage of gas cylinders. The mechanical exhaust of vacuum piping systems is also exempted from the NFPA 99C requirements and must comply with the requirements of the *International Mechanical Code* (IMC).

Gases such as oxygen, nitrous oxide, compressed air, carbon dioxide, helium, nitrogen and various mixtures of these gases are typically used in medical-gas piping systems.

Topic: Design and Installation

Category: Special Piping and Storage

Reference: IPC 1203.1

Subject: Medical Gases

Code Text: *Nonmedical oxygen systems shall be designed and installed in accordance with NFPA 50 and NFPA 51.*

Discussion and Commentary: NFPA 50 applies to bulk oxygen systems that are located at a consumer site and have a storage capacity exceeding 20,000 cubic feet (566 m^3) of liquid or gaseous oxygen, including all unconnected reserves. The bulk oxygen supply originates offsite and is delivered to the premises by mobile delivery equipment. This standard does not apply to oxygen manufacturing plants or oxygen suppliers, or systems having capacities less than 20,000 cubic feet (566 m^3). Either NFPA 51 or NFPA 99 regulates systems less than 20,000 cubic feet (566 m^3).

As a gas, oxygen is odorless, tasteless, colorless and nontoxic, as well as nonflammable. Though oxygen gas is nonflammable, it is an oxidizer and as such creates an environment where the ignition of combustibles may occur more readily.

NFPA applies to oxygen-fuel gas systems for welding, cutting and allied processes; the utilization and storage of fuels; acetylene generation; and calcium carbide storage. The standard does not apply to systems consisting of regulators, hoses, torches, and a single tank each of oxygen and fuel gas. Additionally, systems where oxygen is not used with fuel gases, the manufacture of gases and filling of cylinders, the storage of empty cylinders, and compressed air-fuel systems are not governed by this standard.

Oxygen-fuel gas systems for welding and cutting consist of flammable fuel gases such as acetylene, hydrogen, LP-gas, natural gas, and stabilized methylacetylene propadiene used with oxygen to create a high temperature flame. Although NFPA 51 applies to all fuel gases used with oxygen, very specific regulations governing the generation of acetylene and the storage of calcium carbide are presented because acetylene is the most commonly used gas for welding and cutting.

Study Session 15
IPC Chapters 11 and 12

1. All of the following points of discharge are allowed for the storm water, *except*
   _____ .

   a. lawn

   b. street

   c. combined sewer

   d. sanitary sewer

   Reference _____

2. Which of the following is a prohibited connection in a storm drainage system?

   a. trap

   b. floor drain

   c. quarter bend

   d. backwater valve

   Reference _____

3. Underground storm drainage piping materials are required to comply with _____ .

   a. IPC Table 702.2

   b. IPC Table 702.3

   c. IPC Table 1102.4

   d. IPC Table 1102.5

   Reference _____

4. What part of a storm drainage system does not require the installation of cleanouts in accordance with Chapter 7?

    a. storm sewers   b. subsurface drainage system

    c. above-ground storm drains   d. below-ground storm drains

Reference _____

5. Which of the following materials is approved for above-ground storm drainage pipe?

    a. asbestos-cement pipe

    b. vitrified clay pipe

    c. polyethylene plastic pipe

    d. acrylonitrile butadiene styrene plastic pipe

Reference _____

6. What is the minimum required inlet area of a strainer connected to a 3-inch conductor?

    a. 7.07 square inches   b. 10.6 square inches

    c. 12.57 square inches   d. 14.14 square inches

Reference _____

7. What is the minimum required area of a flat surface type strainer covering a 3-inch conductor, terminating to a drain for a roof parking deck?

    a. 6.0 square inches   b. 9.43 square inches

    c. 14.12 square inches   d. 21.2 square inches

Reference _____

8. What is required between the connection of a storm water conductor and a combined building sewer?

    a. vent   b. trap

    c. strainer   d. backwater valve

Reference _____

9. Where is the cleanout required on a trapped vertical conductor that discharges to a horizontal combined sewer?

    a. on the sewer side of the trap

    b. on the building side of the trap

    c. within 10 pipe diameters downstream of the conductor

    d. a minimum of 10 pipe diameters downstream from the conductor

Reference _____

10. What is the minimum size trap required on a horizontal 4-inch combined sewer receiving the discharge of a 3-inch conductor?

    a. 2-inch   b. 3-inch

    c. 4-inch   d. 5-inch

Reference _____

11. A secondary drainage system is permitted to discharge _____ .

    a. below ground into a subsoil drainage system

    b. above grade, to the ground around the building

    c. into a building sump discharging to a combined sewer

    d. to vertical conductors, below the primary roof drains

Reference _____

12. For a building of 11,500 square feet of projected roof area, what is the minimum required number of roof drains in a controlled flow roof drain system?

    a. one   b. two

    c. three   d. four

Reference _____

13. What is the minimum required size of a semicircular gutter serving a roof with a dimension of 40 feet x 100 feet, located in Duluth, MN?

    a. 6 inches   b. 7 inches

    c. 8 inches   d. 10 inches

Reference _____

14. A secondary roof drainage system is designed to be _____ the size of the primary system.

    a. one-half   b. one and one-half times

    c. two times   d. equal to

Reference _____

15. All of the following piping materials are approved for subsoil drainage systems, *except* _____ .

    a. ABS plastic pipe, ASTM D 2661

    b. PVC, ASTM D 2729

    c. vitrified Clay pipe, ASTM C 4

    d. asbestos-cement pipe ASTM C 508

Reference _____

16. Which of the following is not a requirement of a building subsoil sump pit?

    a. gas-tight cover   b. solid floor

    c. accessible   d. 24-inch minimum depth

Reference _____

17. The maintenance and operation of nonflammable medical gas systems is regulated by which of the following codes?

    a. *International Plumbing Code*® (IPC®)

    b. *International Mechanical Code*® (IMC®)

    c. *International Fuel Gas Code*® (IFGC®)

    d. *International Fire Code*® (IFC®)

Reference _____

18. Roof drains shall have strainers extending not less than _____ inches above the surface of the roof.

    a. 2   b. 3

    c. 4   d. 5

Reference _____

19. A building with a projected roof area of 8,000 square feet located in an area with a rainfall rate of 2 inches would require a vertical leader with a diameter of _____ inches.

    a. 2   b. 3

    c. 4   d. 5

Reference _____

20. A 4-inch horizontal pipe with a 2-percent slope is capable of draining a projected roof area of _____ square feet when located in an area with a 3-inch rainfall.

    a. 1,096   b. 2,506

    c. 3,533   d. 5,010

Reference _____

21. A horizontal projected roof area of 8,500 square feet located in an area with a 2-inch rainfall would require a horizontal pipe of _____ inches when the pipe has a slope of 1 percent.

    a. 3   b. 4

    c. 5   d. 6

Reference _____

22. Semicontinuous discharge from a pump into a building storm drain shall be computed as one gallon per minute being equal to _____ square feet of roof area.

    a. 96   b. 256

    c. 2,000   d. 4,000

Reference _____

23. A 4-inch semicircular gutter with a 4-percent slope in an area with a 1-inch rainfall shall be capable of draining a maximum projected roof area of _____ square feet.

    a. 1,440   b. 2,040

    c. 2,880   d. 4,080

Reference _____

24. Subsoil drains shall be not less than _____ inches in diameter.

     a.   1$^1\!/_2$ b.   2

     c.   3   d.   4

     Reference _____

25. A sump pit shall be not less than _____ inches in diameter.

     a.   18   b.   20

     c.   24   d.   30

     Reference _____

Answer Keys

Study Session 1
2006 *International Plumbing Code*

1.	b	Sec. 101.2
2.	c	Sec. 101.2
3.	c	Sec. 101.2
4.	a	Sec. 101.2
5.	c	Sec. 101.2
6.	d	Sec. 101.3
7.	b	Sec. 102.1
8.	c	Sec. 102.2
9.	d	Sec. 102.3
10.	a	Sec. 102.8
11.	d	Sec. 102.9
12.	d	Sec. 104.2
13.	b	Sec. 105.1
14.	c	Sec. 105.3
15.	b	Sec. 105.2
16.	a	Sec. 105.4.2
17.	b	Sec. 105.5
18.	b	Sec. 102.4
19.	c	Sec. 104.8
20.	a	Sec. 102.6
21.	c	Sec. 104.4
22.	d	Sec. 104.8
23.	a	Sec. 103.2
24.	a	Sec. 102.8
25.	d	Sec. 105.4.3

Study Session 2
2006 *International Plumbing Code*

1.	c	Sec. 106.3.1
2.	b	Secs. 105.4.2, 106.3 and 105.4.4
3.	a	Sec. 107.1
4.	c	Sec. 108.2
5.	d	Sec. 106.2
6.	b	Sec. 106.2
7.	a	Sec. 109.6
8.	d	Sec. 106.5.3
9.	b	Sec. 109.1
10.	d	Sec. 108.4
11.	a	Sec. 107.3.1
12.	d	Sec. 106.5.5
13.	b	Sec. 107.2.1
14.	d	Secs. 109.6, 106.5.4, 104.2 and 104.5
15.	a	Sec. 108.7
16.	b	Chapter 2
17.	d	Chapter 2
18.	b	Sec. 106.5.1
19.	a	Sec. 106.5.6
20.	b	Sec. 109.1
21.	c	Sec. 107.3
22.	a	Sec. 106.3.1
23.	d	Sec. 107.5
24.	d	Sec. 106.5.5
25.	c	Sec. 108.5

Study Session 3

2006 *International Plumbing Code*

1.	a	Sec. 301.3
2.	c	Sec. 301.7
3.	d	Sec. 301.6
4.	c	Table 303.4
5.	c	Sec. 304
6.	b	Sec. 305.6
7.	d	Sec. 306.2
8.	b	Sec. 305.8
9.	b	Sec. 308.5, Table 308.5
10.	b	Sec. 307.5
11.	c	Sec. 307.6
12.	b	Sec. 308.5, 308.6 and 308.7
13.	b	Table 308.5
14.	c	Sec. 310.5, Exception 1
15.	b	Sec. 312.5
16.	c	Sec. 312.9
17.	a	Sec. 314.1
18.	d	Sec. 314.2.3.1
19.	b	Sec. 309.2
20.	d	Sec. 311.1
21.	b	Sec. 307.4
22.	a	Sec. 312.5
23.	d	Sec. 303.4 and Chapter 2
24.	d	Sec. 312.5
25.	b	Sec. 306.2.2

Study Session 4
2006 *International Plumbing Code*

1.	c	Sec. 401.2
2.	d	Sec. 403.1
3.	a	Sec. 403.1
4.	c	Sec. 403.3
5.	b	Table 403.1
6.	d	Sec. 403.4
7.	a	Sec. 403.1.1
8.	c	Sec. 403.2 Exception 2
9.	b	Sec. 405.3.1
10.	c	Sec. 406.3
11.	d	Sec. 406.3
12.	a	Secs. 408.3 and 424.5
13.	c	Sec. 410.2
14.	b	Sec. 410.1
15.	b	Sec. 403.4.3
16.	a	Sec. 405.4
17.	a	Sec. 405.5
18.	d	Sec. 407.2
19.	b	Sec. 405.1
20.	a	Sec. 403.2 Exception 3
21.	d	Sec. 403.4.1
22.	c	Sec. 403.4.2
23.	c	Sec. 405.3.1
24.	b	Sec. 408.3
25.	b	Sec. 411.2

Study Session 5
2006 *International Plumbing Code*

1.	d	Secs. 420.2 and 420.3
2.	a	Sec. 419.2
3.	c	Sec. 418.2
4.	b	Sec. 416.5 and Chapter 2
5.	a	Sec. 417.4.2
6.	b	Sec. 417.4
7.	b	Sec. 421.5
8.	d	Sec. 424.3
9.	c	Sec. 424.3
10.	c	Sec. 414.1
11.	c	Sec. 419.2
12.	b	Sec. 412.3
13.	b	Sec. 412.4
14.	d	Sec. 413.2
15.	a	Sec. 417.5.2
16.	c	Sec. 425.2
17.	b	Sec. 425.3.2
18.	a	Sec. 422.6
19.	b	Sec. 425.1.1
20.	b	Sec. 425.3.1
21.	c	Sec. 422.9.3
22.	b	Sec. 416.3
23.	a	Sec. 417.3
24.	d	Sec. 421.5
25.	c	Sec. 424.5

Study Session 6
2006 *International Plumbing Code*

1.	d	Secs. 501.3, 503.1 and 504.1
2.	c	Sec. 504.2
3.	d	Sec. 504.5
4.	c	Sec. 504.4.1
5.	d	Sec. 504.5
6.	c	Sec. 504.7
7.	a	Sec. 504.7.1
8.	d	Sec. 504.6
9.	c	Sec. 504.6
10.	a	Sec. 504.5
11.	d	Sec. 504.6
12.	c	Sec. 505.1
13.	c	Sec. 504.7.1
14.	b	Sec. 501.6
15.	b	Sec. 503.1
16.	b	Sec. 501.5
17.	c	Sec. 501.2
18.	c	Sec. 501.7
19.	a	Sec. 502.3
20.	d	Sec. 504.3
21.	a	Sec. 504.7.1
22.	d	Sec. 504.7.2
23.	c	Sec. 504.4.1
24.	b	Sec. 501.2
25.	b	Sec. 502.3

Study Session 7
2006 *International Plumbing Code*

1.	b	Sec. 602.3.1
2.	c	Sec. 604.8
3.	b	Sec. 604.8
4.	c	Sec. 604.9
5.	b	Sec. 604.9
6.	d	Sec. 605.2
7.	a	Sec. 604.8
8.	d	Table 605.4
9.	a	Sec. 605.4
10.	b	Secs. 605.3, 605.4, 605.5 and 605.7
11.	b	Sec. 605.24.1
12.	d	Secs. 605.16.3 and 605.22.3
13.	c	Sec. 604.6
14.	c	Sec. 605.3
15.	b	Table 604.5, Footnote a
16.	b	Secs. 605.14.3 and 1605.15.4
17.	b	Table 604.4
18.	b	Sec. 604.4, Exception 1
19.	a	Sec. 602.3.5.1
20.	a	Sec. 603.1
21.	c	Sec. 603.2
22.	d	Sec. 603.2, Exception 2
23.	c	Sec. 605.16.2, Exception
24.	a	Sec. 605.16.2, Exception
25.	c	Table 604.5

Study Session 8
2006 *International Plumbing Code*

1.	d	Sec. 606.2
2.	d	Sec. 606.4
3.	c	Sec. 607.1
4.	b	Sec. 606.5.4
5.	c	Sec. 607.3.2
6.	b	Table 608.1
7.	d	Sec. 608.16.4.1
8.	d	Sec. 607.4, Exception
9.	a	Sec. 608.16.1 and 608.16.10
10.	d	Sec. 608.16.5 and Table 608.1
11.	d	Sec. 608.8.1
12.	c	Table 608.15.1
13.	a	Sec. 608.16.3 and Chapter 2
14.	d	Sec. 608.17.6 and Chapter 2
15.	b	Sec. 610.1
16.	c	Sec. 608.3.1
17.	c	Sec. 606.5.9
18.	b	Sec. 608.13.4
19.	c	Sec. 608.13.1
20.	c	Sec. 608.8
21.	c	Sec. 608.16.1
22.	c	Sec. 607.2
23.	c	Sec. 611.1 and Chapter 13
24.	b	Sec. 612.1
25.	a	Sec. 606.5.1

Study Session 9
2006 *International Plumbing Code*

1.	d	Sec. 701.3
2.	c	Sec. 701.7
3.	d	Sec. 701.9
4.	b	Tables 702.1 and 702.2
5.	c	Sec. 702.5
6.	d	Sec. 705.16.2
7.	a	Sec. 705.14.2
8.	b	Sec. 705.18.2
9.	c	Sec. 705.18.7
10.	d	Sec. 705.14
11.	c	Sec. 706.3
12.	b	Table 706.3
13.	d	Sec. 706.4
14.	c	Table 704.1
15.	a	Sec. 704.3
16.	a	Sec. 703.3
17.	a	Sec. 704.2
18.	d	Sec. 704.5
19.	a	Sec. 705.2.2
20.	c	Sec. 705.14.1
21.	b	Sec. 705.18.4
22.	a	Sec. 705.20
23.	c	Sec. 706.2
24.	b	Table 706.3
25.	d	Sec. 706.3

Study Session 10
2006 *International Plumbing Code*

1.	a	Sec. 707.1
2.	b	Sec. 708.3.3
3.	c	Sec. 708.7
4.	d	Sec. 709.2 and Table 709.2
5.	c	Sec. 709.3
6.	a	Table 709.1 and Chapter 2
7.	a	Table 710.1(1), Footnote a
8.	b	Chapter 2
9.	d	Sec. 712.1
10.	a	Sec. 712.1 and 712.2
11.	b	Sec. 712.3.4
12.	c	Sec. 712.3.2
13.	d	Sec. 712.3.5
14.	c	Sec. 712.4.2, Exception 1
15.	b	Table 712.4.2
16.	c	Sec. 713.9
17.	c	Sec. 713.3
18.	c	Sec. 714.3.2 and Table 704.1
19.	b	Sec. 715.4
20.	a	Sec. 715.1
21.	b	Sec. 713.7.2
22.	d	Sec. 708.3.1
23.	a	Sec. 708.3.3
24.	c	Sec. 708.3.4
25.	b	Sec. 708.8

Study Session 11
2006 *International Plumbing Code*

1.	a	Sec. 802.1.1
2.	c	Sec. 802.1.7
3.	c	Sec. 802.1.2
4.	c	Sec. 802.2
5.	b	Sec. 802.3.1 and Table 710.1(2)
6.	b	Sec. 802.3.2
7.	c	Sec. 803.2
8.	b	Sec. 803.1
9.	a	Sec. 803.3
10.	d	Sec. 802.4
11.	a	Sec. 804.1
12.	b	Sec. 802.3
13.	d	Sec. 802.1.4
14.	c	Sec. 802.2
15.	d	Sec. 802.4
16.	b	Sec. 802.1.6
17.	a	Sec. 802.3
18.	d	Sec. 802.3.2
19.	b	Sec. 802.2.1
20.	b	Sec. 801.2
21.	a	Sec. 802.1
22.	b	Sec. 802.1.2
23.	b	Sec. 802.3.1
24.	d	Sec. 803.2
25.	a	Sec. 803.1

Study Session 12
2006 *International Plumbing Code*

1.	b	Sec. 901.2
2.	a	Sec. 901.2
3.	c	Sec. 901.3
4.	b	Secs. 902.2 and 902.3
5.	a	Sec. 903.1
6.	c	Sec. 903.4
7.	b	Sec. 904.2 and Appendix D
8.	c	Sec. 904.2 and Appendix D
9.	c	Sec. 904.1
10.	b	Sec. 904.5
11.	d	Sec. 904.6
12.	d	Sec. 904.7 and Appendix D
13.	d	Sec. 905.4
14.	c	Sec. 906.1 and Table 906.1
15.	d	Sec. 906.2
16.	d	Sec. 906.1, Exception
17.	b	Sec. 908.1 and Chapter 2
18.	b	Sec. 908.3 and Table 908.3
19.	c	Sec. 907.1
20.	b	Secs. 909.1 and 909.1.1
21.	a	Secs. 909.1 and 909.1.1
22.	b	Sec. 909.1
23.	b	Sec. 909.2
24.	d	Table 906.1
25.	a	Sec. 906.3

Study Session 13
2006 *International Plumbing Code*

1.	d	Sec. 910.3
2.	d	Table 910.4
3.	d	Table 910.4
4.	b	Secs. 910.2 and 910.4
5.	c	Sec. 913.2
6.	a	Sec. 916.1
7.	d	Sec. 916.4.1
8.	a	Sec. 911.4
9.	b	Sec. 911.4.2
10.	a	Sec. 912.1
11.	c	Sec. 912.2.1
12.	c	Sec. 912.2
13.	c	Sec. 912.2.4 and Table 906.1
14.	b	Table 912.3
15.	b	Sec. 913.1
16.	b	Secs. 913.3 and 916.2
17.	d	Sec. 912.1
18.	d	Sec. 914.1
19.	b	Sec. 914.2
20.	d	Sec. 915.2
21.	d	Sec. 916.2
22.	c	Table 916.5.1
23.	a	Secs. 917.3.3 and 917.8
24.	d	Sec. 713.2
25.	b	Sec. 917.4

Study Session 14
2006 *International Plumbing Code*

1.	b	Sec. 1002.1
2.	c	Sec. 1002.3
3.	d	Sec. 1002.10
4.	b	Secs. 1002.2 and 1002.4
5.	a	Sec. 1002.3
6.	b	Sec. 1002.5 and Table 709.1
7.	d	Sec. 1002.6
8.	b	Sec. 1002.1
9.	a	Sec. 1002.2
10.	a	Sec. 1002.1
11.	d	Sec. 1003.1
12.	b	Secs. 1003.2 and 1003.3.1
13.	c	Sec. 1003.3.2
14.	d	Sec. 1003.3.3
15.	c	Sec. 1003.4.2.2
16.	a	Sec. 1003.5
17.	c	Sec. 1003.3.4.2
18.	b	Sec. 1003.3.5
19.	d	Sec. 1003.9
20.	b	Sec. 1002.1
21.	c	Sec. 1002.4
22.	d	Sec. 1002.9
23.	c	Table 1003.3.4.1
24.	a	Sec. 1003.4.2.1
25.	c	Sec. 1003.4.2.1

Study Session 15
2006 *International Plumbing Code*

1.	d	Sec. 1101.2 and Chapter 2
2.	b	Sec. 1104.3
3.	a	Sec. 1102.3
4.	b	Sec. 1101.8 and Chapter 2
5.	d	Sec. 1102.2 and Table 702.1
6.	b	Sec. 1105.1
7.	c	Sec. 1105.2
8.	b	Sec. 1103.1
9.	b	Sec. 1103.4
10.	c	Sec. 1103.3
11.	b	Sec. 1107.2
12.	d	Sec. 1110.4
13.	b	Table 1106.6 and Appendix B
14.	d	Sec. 1107.3
15.	a	Table 1102.5
16.	a	Secs. 1111.1 and 1113.1.2
17.	d	Sec. 1201.1
18.	c	Sec. 1105.1
19.	c	Table 1106.2
20.	c	Table 1106.3
21.	d	Table 1106.3
22.	a	Sec. 1109.1
23.	d	Table 1106.6
24.	d	Sec. 1111.1
25.	a	Sec. 1113.1.2